Speaking Computer
Learning the foreign language of technology

by Scott Shui

Speaking Computer
Learning the foreign language of technology

by Scott Shui

Alephsphere Publishing
Cyberspace

North American Edition 2012
Version 0.90 (beta release)

For information about special discounts for bulk purchases, please contact Special Sales at   suntzu [at] alephsphere.com

ISBN 978-0-9852-9010-8

"Any sufficiently advanced technology is indistinguishable from magic. "

-Arthur Clarke

Dedicated to Mom and Dad.

Inspired by Karen Elizabeth Gordon and Bill & Melinda.

Contents

Foreword

* Watch this space.

*   Ideally I want Shakira to write it. Why? <u>Speaking Computer</u> is written for explorers, dreamers, and survivors. It is a technology book for those who can appreciate the wonder of computers but are busy living their lives. This work is here to tame the restless, comfort the cautious, and entertain the brave.

To appeal to everyone in the audience, we need to hear from a calming voice outside the realm of machines. She should embody a sense of emotional empathy to reach people who might be wary of computers. Of course she should also be intelligent, approachable, and well traveled enough to interact with anyone. If she can grasp the edges of the human spirit then she can understands the wondrous potential of the internet.

As a multilingual singer and songwriter, Shakira embraces the passions of the soul. As a scholar and humanitarian, she knows the precious value of life and the mind within. As an artist who strives to break creative boundaries, her boldness encourages us to explore something as unfamiliar as an engineered world. She is a role model for each of us facing our own complex challenges.

Normally a book has its foreword written by someone with a industry credentials in order to provide a recommendation of relevant experience. To reach a truly diverse audience <u>Speaking Computer</u> has to be endorsed by a modern renaissance man or woman. The testimonial has to come from a dynamic individual like Shakira.

Let her know.

Introductions

My little work of words rips open the computer universe for you to see all of its warm and fleshy insides. The guts of the machine will be dissected with mental scalpels to reveal the chips, history, logistics, psychology, and puppet strings. We will discuss the perpetually developing landscape of software and the ever evolving outer shell of a personal computer. This exposition will lay bare the faux soul of technology.

For the millions of people who do not speak Computer, I am here to translate it to both sides of the brain. With some imagery from the real world, the abstract tech stuff will make a lot more sense. Your laptop has a brain kind of like you do. Instead of blood and arteries, it has electricity and wires. Its DNA is written in a binary language of data pulses rather than a double helix of microscopic nucleotide pairs. Pretty much every aspect of the electronic universe has a vivid visual to ease our comprehension.

This is for the restless artists, parents from another era, angst filled technophobes, and casual users who would normally walk past a computer book on the shelf. To defray your fears of a wired planet, I put the big bad machine inside a glass cage while we stop and stare. From an arm's length we poke at the keys to reveal some of the dangers of an unrestrained internet enabled computer. A quick peek at the tools it provides should be enough to show how it is indispensable as a business partner. It may beep and bark like your typical pet but it can play games, sing to you, and obey commands better than any puppy dog. The feel of foreignness will lessen over time as you get to know each of its subtle quirks and nuances.

Advanced coders and hardware folks might give me flack for exclusions or nit pit about my choice of imagery but please go easy on the amateur programmer who could not hack it. Once upon a time, I did stare at Unix terminals for hours at a time fighting my compiler...but I am feeling much better now. The flashbacks are becoming less frequent and the Klingon Tourette's has greatly subsided.

As I started to mention before, this computer book is for you mom. The fact that you and dad raised all of us without emailing anyone for help or looking up anything on the web is a miracle by today's standards. All the yummy recipes you had came from some mysterious maternal place and not an online forum. Your stories originated from the life you led and the people you met, not from a string of email forwards. Do you remember a few years ago when you pulled me aside and asked me to teach you how to use the VCR? Whenever you are ready to tackle the internet, grab a copy of my book and let me know.

This applies to you too dad. Although I have never seen you step within reach of a keyboard, the excuse about your age does not work. From my oldest memory to my most recent one, you are always reading a newspaper or watching the news; besides the local paper and the nearest big city one thrown on our porch, you also get a couple international ones every day in the mailbox. If you can figure out what is going on in the world, it is no big deal to turn on a computer to surf the internet. When one of those big computer companies get around to making a thin, light, and large internet enabled touch tablet, you are totally out of excuses.

To all the people left behind in the technological rapture of the computer age, this book is for you. It is your official invitation to read up on what you have been missing. For the price of admission you get the basic concepts, a few tools, a teaser about the toys, some special features, and a brief glimpse of the future. Despite all the details packed into these pages, this is still just a starting point for your trip. This primer will not make you an expert but it should give you a place to stand.

These days, basking in a blissful ignorance can put you at a distinct disadvantage. Your worst fears about how complicated computers are pale in comparison to living la vida Luddite. Most financial groups and government agencies are rapidly becoming so automated that online accessibility typically grants you a faster response. A simple call to customer service in many industries might take an hour or more while live web chat support takes seconds to respond. Countless business and consumer resources, which used to require intense hours of research in a large metropolitan library, are accessible now in mere minutes on the internet. In fact, the equivalent of a university education in dozens of fields is freely available to any online scholar ambitious enough to do some independent study.

What does it take though to bring the average computer user up to a functional proficiency? For someone to be considered literate, we dictate one's reading skills to be at a certain level by measuring his vocabulary and comprehension. In the case of modern technology, we associate a person's basic aptitude with the ability to identify the latest device, to learn new programs, and to troubleshoot simple problems. Anything less than that means your computer will either be vastly underutilized or you will be a in constant struggle to fix things.

To go beyond the fundamentals requires an intense dedication of time and energy, as well a reasonable degree of intelligence. Delving into the structural code of any piece of software is akin to dissecting a biology specimen. There is an underlying logic in the way it works as well as an art to the way it looks. Working with computers may seem complex at first but it is essentially just another craft to master. Rising from an amateur dabbler to a professional craftsman requires as much as any other trade. Be sure to pay attention to all the artists, scholars, frauds, gurus, and apprentices you meet as you ramp up your technical prowess.

We are all on the way to improving our expertise with humanity's most useful and universal invention since the wheel. Until we create a magic information pill, we are still going to have to learn things the old fashioned way. A little bit of watching tutorials, a lot of reading, maybe some audio lessons, and perhaps a few thousand hours of actual hands on time to tinker, touch, and explore.

Instead of dragging you into a closed classroom and forcing you to memorize step by step instructions on "how to this" and "how to that," the focus is going to be on what you already know. By borrowing from pieces of life, we can talk about computers in a way that is more real, literally closer to home. The act of staring at a flat glass screen is as lifeless as you can get unless we imagine the abstract insides as tangible objects we can hold, throw, see, feel, or break. Our brains happen to like the solid sense of a world we can interact with by touch, sound, and sight.

To picture the files on your computer desktop metaphorically as stacks of paper and nested manila folders, we use one of the time honored traditions of storytelling. Describing the truly unfamiliar in terms we can understand helps to solidify ideas in our minds. This bridging creates a frame of reference for anyone not raised on the concepts; obviously, an ideal education would have technology thoughtfully integrated into the curriculum. A lot of times we turn to metaphors to explain things we have never seen before. Even when we know a concept inside and out, conveying a new idea to someone can be challenging. As accurate as a concise definition may be, our wiry minds tend to remember strong imagery over abstract thoughts.

Amusing alliterations, popular culture references, doled out clichés, personal narratives, and outright dramatic flourish also serve to make the computer commentary more colorful. Amidst the flashy words, there is a lot of information crammed into this book. The density of data poured into this splatter of ink should be enough to keep you enthralled, entertained, and hopefully enlightened.

Without this splash of literary spices, a book about computers is dull and dry to a normal person. It lacks the sensual fantasy of a romance novel and the gripping thrills of a taut murder mystery. There will not be any intrepid adolescent boy wizard, nine panel illustrations, sexy vampires, secret transformation tips, or food strategies to draw you in. It will not be found in the over 18 adult oriented section or surrounded by children's books either; those would be some peculiar computer books though. Unless you have some big technical goals, the likelihood of you picking up a programming book just for fun is pretty darn low.

At first glance, the computer itself has all the charisma of a corpse. It sits there motionless on the table or desk waiting for you to do something. That hunk of plastic and glass usually has the beauty of a wooden log and the fragility of your grandmother's fine china; there is a popular fruit themed one encased in sleek metal too if you so desire. It costs hundreds, if not thousands of dollars, and becomes obsolete in just a couple of years. After a lifetime of

4

studying, you will never completely understand how everything works with it. Of course, if you had a nickel for each time something went wrong or you encountered a system error, you would be beyond rich.

Would you know that this humble electric thing is a work of art, science, and industry? On a good day, it allows us to individually manage large corporations, play with bundles of molecules, or bring cinematic worlds to life. Whatever we can grasp by the power of our minds, technology will provide us the reach.

So check it out.

* * *

"I know you're out there. I can feel you now. I know that you're afraid... you're afraid of us. You're afraid of change. I don't know the future. I didn't come here to tell you how this is going to end. I came here to tell you how it's going to begin. I'm going to hang up this phone, and then I'm going to show these people what you don't want them to see. I'm going to show them a world without you. A world without rules and controls, without borders or boundaries. A world where anything is possible. Where we go from there is a choice I leave to you."

-Thomas Anderson, aka Neo, from the movie The Matrix

Anatomy of the Machine

1

There is no blood running through the veins of a computer. It does not cry when you scream at it in anger or curse obscenities at it in frustration. We bargain in vain with it to give us what we want, like prisoners to a Carribean jailer's dog. By casually anthropomorphizing this television typewriter combination into a person, we trigger science fiction's wet dream of a surrogate plaything. However it does not care, think, have feelings, or possesses any kind of awareness...at least not yet.

A computer is best described metaphorically to anyone whose fascinations lie outside the meticulous world of robotic toys or the linguistics of C++. For a person immersed in the world of computers, a "Blue Screen of Death" is a simple historical reference to a major system error. To anyone else, a BSOD just sounds like a horribly bad condition which lacks an everyday frame of reference to make sense. Experiencing one is like stepping on the proverbial rusty nail. They both happen suddenly and hurt a great deal. The trip to an emergency room for a fix has promises of necessary pain, either financial or physical.

Compared to how a human being works, a computer is far simpler creature. Your trusty laptop/desktop just needs a constant supply of electricity to stay alive and awake. Occasionally you may want to take some time to hand detail your dusty desktop and dirty monitor like a spoiled car. However, a person's incessant need for clean water, nutritious food, and social interaction causes him to be very high maintenance. While it is true that everyone poops, a Dell will not lay a dookie.

Inside the case of every personal computer is the CPU, the central processing unit, the brains of the outfit. Each generation that is born from processed sand(silicon) and soda cans(aluminum) is created to run faster than last year's model. Under the laboratory microscope it is a flat little piece of glass with microscopic streets. When it leaves the factory it has plastic and metal packaging that makes it look like a miniature chocolate bar with silver spider legs.

Intel and AMD are the two major companies that make CPUs for the personal computer market, locked in a perpetual Coke and Pepsi style brawl; ironically they continually compete to release a faster processor for bragging rights when both can effortlessly handle most people's obsessive need for email and web surfing. In my travels I have watched the speed increase an order of magnitude from Megahertz(MHz) to Gigahertz(GHz), faithfully following Moore's Law over the years; in 1965 Gordon Moore predicted the complexity of an integrated circuit would double roughly every two years. A processor speed limit does exist at the quantum level when making atomic sized transistors; the electrons start behaving badly when you don't give them enough room to move. Today's trend is to combine processors into groups of two and four or more to increase the overall power of the artificial brain; we can expect the next generation of personal computer have to have a hive mind cpu.

The second most important type of chips in any computer is the RAM(random access memory) which determines how many programs, processes, and files you can juggle at a time. The more RAM you have, the more you can manage without the system slowing down to a crawl. Think of it as how many things you can carry in your hands at once, whether it is a chainsaw, a pair of scissors, or a cup of hot coffee filled to the brim. Adding more RAM to a computer is like outpatient surgery in that it is usually immediately beneficial, there are limits to how much it can help, and relatively painless and quick as long as you know what you are doing.

As the dominant storage system in today's personal computer, the hard drive is currently measured in hundreds of Gigabytes(GBs). A few laptops being sold today have less than that just as a more expensive desktop machine may have a terabyte(1TB= 1000 GB) or more. Your hard drive is your computer's special backpack that holds things for you when you start your computing hike. Inside the metal shell is a magnetic disc spinning at several thousand revolutions per minute(RPM) with a needle that reads the data off of it like a record player(remember those?) with a hovering arm; the higher the RPM, the faster you can access information.. Futurists are already saying that these drives will be replaced in a

8

few years by solid state drives(SSDs), larger versions of those memory chips used in digital cameras; the data stays in there when the power is off and it has no internal moving parts.

Inelegant but necessary, the power supply is the heart of the machine, directing power from the wall to the entire system. It works quietly and rarely breaks down but when it does, you know it immediately by the blank screen and silence. In a laptop it sits encased in a thick sheath of hard plastic as the warm brick inline to your power cord; this juices up certain batteries to eight hours of life at a time... sadly most notebook configurations will only let you last long enough to watch a movie.

The motherboard is the main circuit board onto which all components connect centrally. It is the backbone, the spine, the central nervous system of the computer. Sharp plastic cards embedded with microchips are plugged tightly into narrow sockets like axes stuck in a tree trunk. DVD burners and hard drives are tethered by bands of insulated wire to the main board and power supply, mimicking nerves and arteries. Every port on the outside of the laptop or desktop case carries data as two way electrical streams to distinct parts on the motherboard.

On the laptop I am using for the first draft of *Speaking Computer*, I count over eight types of slots, ports, and metal lined plastic holes where a playful child can jam with peanut butter. One trapezoid(remember your geometry) shaped jack(mine has the old analog style with three rows but the newer digital type is larger with two rows) lets you connect to a monitor and a round one that looks like a mouth with too many eyes lets you watch your computer on TV(S-video). Another one accepts the fairly familiar phone cord while a similar yet wider one helps you connect to one or more computers, creating a network; the latter is an ethernet jack whose name sounds as if you are hooking up to a virtual world of empty air. A pair of audio ports allow you to plug in a microphone and headphones; it's the small round hole like on your Walkman, I mean iPod. A thin slit with the letters SD, MMC, and or Xd indicate a place for the memory card from your new digital camera. Several rectangular slots about a half inch wide called USB ports have the versatility to communicate interchangeably with any number of devices including mp3 players, cell phones, printers, scanners, hard drives, mice, keyboards, tablets, flash drives, flashlights, fans, wireless transmitters, joysticks, massagers, cup warmers, pencil sharpeners, and even mini lava lamps. Conveniently and perhaps confusingly, each socket also has an engraved or printed symbol that confirms its function if you knew what it meant in the first place.

Newsflash on the replacement laptop I am using to edit the final version you are reading. The modem jack has disappeared since no one uses landlines anymore and so has the s-video port, not that you noticed. An HDMI port also now graces the side of my fancy schmancy machine; with a handy cable my laptop can now make sweet loving to any fast and loose big screen TV. Are you checking email on your TV yet?

Until voice recognition programming is perfected and widespread, there will still be an alphabet based keyboard to convey our commands; my crazy new Windows machine does have a microphone and the program included as part of the system but I still type faster than I can talk. The familiar QWERTY sequence is a holdover from when typewriters ruled the office and was designed to prevent the keys from getting stuck. A manufacturer can switch over to a regular A to Z configuration easily but there would be the matter of weaning entire generations of people off the legacy effect. For fun and ergonomics, you can find keyboards split into halves for gaming overlays or contoured into small mounds of grouped buttons.

Based on optimistic science fiction and actual advances in manufacturing, we will inevitably all use touchscreens and/or holographic air displays; the interactive imaging in the movie Minority Report(2002) is just not cheap enough for everyone yet. Sure the graphic professionals have highly sensitive tablets that work like remote controlled drawing pads; laptop users have a smaller built in version called a touchpad . Hardcore Centipede fans and office workers with sensitive wrists will choose to go with a track ball controller(rolling a billiard ball mounted on a plastic stand *is* a fun way to kill spiders). On some laptops you may also find a pointing stick situated in the midst of the keyboard mimicking a misplaced eraser from a #2 pencil. For the overwhelming majority, we have to make do with a device about the size of a bar of soap called a mouse. On the underbelly of the mouse you will usually find either a rubber ball or a laser sensor to track planar movement and feed it back to the corresponding place on the monitor.

The monitor has an evolution tracing back to the early days of black and white television. Essentially they are one and the same except a TV has a tuner so it can interpret broadcast signals from the atmosphere. As long as you have the right cable and connectors, they are virtually interchangeable in theory. During the start of the personal computer age, you did plug the computer into the boob tube; come on kids, ask the old people about this one. As personal computers became more advanced, they also started making computer monitors that surpassed the resolution of your home television. Every image you see is measured in dpi(dots per inch) and with the screen set to a given resolution it appears as sharp and small or large and fuzzy; web browsers are optimized for pictures at 72 dpi although printouts are exceptionally sharp at 300 dpi and higher.

A printer is the magical machine that you put blank paper in and out comes pages of pictures, love letters, coupons, birthday cards, family recipes, psalms, ransom notes, dirty jokes, and customized maps spewing forth. The three major types of pretty paper makers in order of increasing cost are ink jet, laser, and dye sublimation. One consumes ink based on the three primary paint colors and our preference for simple black print. Unlike an uzi that litters spent shells on the ground, an ink jet printer simply gives you annoying messages to change out its cartridges. Using a large heated metal roller, an internal laser, and a whole lot of pixie dust the laser printer is the fastest image creator of the bunch; for you muggles, dry toner powder in the same 3+ 1 color matrix is used instead of crushed diamonds mined by dwarfs. Dye sublimation is a Mr. Wizard way of having primary color screens of dye heated one by one onto a surface so that the color bleeds onto it as a gas thus making a layered image; think of it as big yellow, red and blue spotlights with cutout masks converging together.

Acting as the voyeuristic stepbrother is the scanner, a specialized high resolution device that reads a sheet of paper inch by inch from top to bottom. Sometimes it is hybridized with a printer so that you have a copier. Primarily it is an archival tool that creates a electronic version of whatever you have in paper form one page at a time. This invention is the key to transforming the idea of a paperless office into reality but human beings love the tactile nature of paper too much for that to happen.

Although we are getting closer to the day where every computer is perpetually and universally connected, we still have a primitive tendency to want to physically carry data around, to hoard archives like treasure, and treat the tome itself as sacred. Information storage is evolving from paintings on cave walls, to papyrus scrolls and bound paper books, to carefully engineered talismans encoded in a simple yes or no digital language. Magnetic media in the form of cassette tapes, floppy disks, and hard disc drives are becoming obsolete in descending order of imminent extinction. Optical discs made from sandwiched layers of metal and clear plastic have the potential to last through your grandchildren's lifetime even as the capacity multiplies for each generation; advances in lasers and layering techniques in manufacturing means that 1 blu ray disc(50GB double layer) = 71 CDs and one CD can hold hundreds of books; a regular single sided DVD holds about 4.5GB compared to the 0.7 GB(the same as 700MB) on a CD. Researchers have already shown in a laboratory that we can access information in protein gel by shining a laser at the molecular level; Alice's next trip through the looking glass may one day bring back libraries we can swallow.

Sometimes we are so mesmerized by the bright colorful screen we often ignore the computer's audio potential until we plug in a cheap pair of speakers. Playing a music CD while you toil away in your cubicle is its mindless menial day job. By streaming music off the internet you are enabling a radio with the every-single-station-in-the-world feature. For garage band duty, the basic hardware is enough to be a mixer, a sound board, a special effects generator, a recorder, an editor, and a speech synthesizer. Professionals and technically minded musicians will install a sound card with chips specifically dedicated to processing the audio algorithms.

Suppose you want to fulfill the Jetsons' prophecy of a

two way video phone or better yet compete head to head with network television. You can hook up the starving student's setup of a small microphone and a web cam(short for camera of course) in a few minutes and converse with someone on the other side of the world; if you have a relatively new laptop, it is already built in. Monologue your deepest darkest secrets, hit the record button, and post it to a popular web site. Public access television is now available without a trip to the local cable TV studio, only now the audience is worldwide. Spend a few hundred bucks on a good digital video camera and add a tripod to the mix and you can change Nielsen's paradigm.

The current generation of mobile computers have a built in wireless feature that lets you access an internet connection anywhere in your house using a device called a router that broadcasts data to you. Wifi as it is it commonly called has a range that can extend to your neighbor's yard or easily from a dozen sources in an apartment complex. An older computer can do this by plugging in an adapter in the form of PCI card for a desktop, PCMCIA card for a laptop, or a USB dongle for either. Usually the signal is encrypted or locked so that you need a password, numeric key, or passphrase to connect. An unlocked signal is a check-your-email and surf-the-web freebie offered by many coffee shops luring you in to sample their caffeinated wares. A longer range alternative that utilizes the cellular phone network is starting to gain ground but so far it is just an optional add on; if you have the unlimited data plan, you can just wirelessly tether it to your Android or iPhone too.

For those those skateboarding along the information superhighway, dialing into a internet service provider is still a way of life. A simple phone cord connects the computer's built in modem directly with the phone outlet in the wall. The prolonged stuttering beeps followed by the sounds of static over the line is still familiar to the millions who have not yet migrated to a broadband connection. Even as the usage of the modem is diminishing, it still can be used to send and receive faxes, another relic of technology; if you do not know what a fax is, that is okay since you can learn about it later on the History Channel.

At home or at any company without a full time

IT(Information Technology is the fancy name for the computer people) department, the high speed hookup to the internet will usually be by DSL, cable, satellite, or FIOS. DSL(means Digital Subscriber Line so you can answer that middle row in the crossword puzzle) is handled by the phone company or partner to the phone company and utilizes a DSL modem on your existing phone line. Your cable TV provider uses the same carrier signal to send data to a cable modem box. Anyone far from civilization and possessing generous funding can take advantage of internet via space using a dish and a satellite modem. FIOS or Fiber Optic Service uses the same fundamental technology as the backbone or the internet itself, light traveling through semi-flexible insulated glass thread. This always requires a technician to come by and install a Optical Network Terminal on the side of the building to connect to your computer or router.

Although routers were briefly mentioned before as a wifi source, it does a couple other very important things. It enables you to connect multiple computers so they can all access the internet and each other. It also hooks up your video game system, Tivo or generic DVR(digital video recorder that records your soap operas to a hard drive), or PDA(personal data assistant that is wifi enabled). Perhaps its most valuable attribute is that it generally has a firewall hard wired into the circuitry that hinders hackers from prying into your computer. That electrically powered plastic box which we take for granted is the one with the antenna and several ethernet ports(the ones that look like phone jacks but bigger) on the back; it may have the word Netgear, Linksys, D-Link, Cisco, Buffalo, or Airlink branded on it.

If you value your equipment at all, you will have every one of your devices connected to a quality surge protector in case of blackouts and fluctuations in electricity, which by the way occur completely out of your control. An outlet multiplier or a simple power strip(this does not have a circuit breaker on it) is just as bad as plugging it directly into the electrical socket in the wall. Anyone who I catch doing this will have to hand write, "I will always use a surge protector for my computer equipment," five hundred times. In cursive.

This is where I give an open letter of thanks to all those generous businesses who offer free wireless internet and allow customers to plug into your electrical outlets. Some people take it as an automatic privilege to use your electricity in exchange for purchasing that tall latte. They seek out that discreet corner with the comfy chair and entrench themselves in for that marathon session of browsing the net for news, writing papers, crafting emails for a startup company, making phone calls overseas, updating their daily blog on why the world is doomed, or just watching their favorite TV shows. It is like the neighbor's kid that comes over to mow your lawn for free and ends up staying all afternoon watching TV at your house. Hopefully you are not too bothered by the electric bill suddenly jumping up a hundred dollars each month because of all the slackers, I mean nouveau entrepreneurs, hanging out and stretching that cold cup of java over three, four or five hours. So if there is a power surge that damages all those leeching laptops, it would be a case of electricity karma, right?

To extend the portability of my power dilemma, I bought an extra battery that I keep with me fully charged so that I can spend a full work day staring at my notebook without being tethered. Any longer than that is probably more than I want to be without surfacing for air, food, or extraneous movement. As amazing as this invention is, I am lucky enough to be able to take a step back and see what effect this computer thingy has on my anatomy.

***

The pandemic of the modern workplace involves hundreds upon hundred of hours a year in a sitting position in front of a fixed computer monitor or two. A convicted felon has more freedom of movement in his cell than many of today's wage earners. For the average person this unnatural habitat is a nightmare of ergonomics which takes its toll on every part of the human body. Working inside an air conditioned office seems like it would be safer than the obvious hazards of being a construction worker, fireman, or assembly line technician but the subtle effects of a cubicle are silently cumulative. Your mental health is also at risk because like those poor animals at the zoo, a certain amount of atrophy affects the spirit of something trapped in a cage.

Pretty much most kids learn from their parents, teachers, and personal experience that staring at the sun is a bad idea. Those lingering spots are momentary damage to your eyes that become permanent if you make a habit of it. Halogen light bulbs, photography flashes, and strong reflections off mirrored surfaces leave a similar after effect. Watching TV or using a computer is deceptively safer because you can stare at them for hours at a time before eyestrain sets in. We forget that we are gazing at a rectangular light source as well as a fixed plane in space. We forget how day after day of this habit affects our eyesight until we need to go see the optometrist to change our prescription.

If you have made your work life such that you are fated to spend it programming code, writing articles, operating repetitive accounting software, or otherwise parked in from of the glowing screen, you need to take care of your eyes. Frequent breaks outside of your office will give your eyes the opportunity to focus on different points in space and to rest from staring at the small words on the face of the screen. A hawk in flight is able to spot its prey while it is moving because it is constantly scanning its entire field of vision. Being in that stationary chair is isolating your field of vision to a small area of space, forcing you to have an anti-social stare as an unwanted byproduct. Any serious imbalance between the brightness of the screen and the light in your office is bad for you too. When your screen's relative brightness is really high, you probably notice your eyes getting tired after a few hours rather than the end of the day. If your screen is too dim then you will notice that it takes longer to focus on objects away from the monitor. A spot check to adjust the brightness to a comfortable level takes a few minutes but most likely, you will leave it alone because it has always been like that way. There is probably an expensive device that you can use to calibrate the optimal brightness relative to the ambient light in the room but a simple rule of thumb should do fine. <u>Your monitor should be as bright as a page from your favorite magazine that you read lounging in your spare time.</u> The lights in your office should also provide enough visibility to make your screen neutral to its surroundings, not a dark page, not a burning white, and not a glaring mirror. If your eyes feel weary at the end of the day, there is more fine tuning to do for tomorrow. It may take several days to get your monitor settings and room lights in balance but it will be worth it in the long run.

Preserving the integrity of your neck and back is something of major earth shattering importance too. Written in the fine print of your job contract, beneath the paragraph about nuking leftovers in the microwave for lunch, is the stipulation that you will have to wear this invisible harness immobilizing your spine ten odd hours a day. Cradled in your six hundred dollar office chair for a full work shift plus another few hours commuting on the freeway is exactly how you imagined life as a grown up. Strangely enough that high school career counselor neglected to mention how stiff you would feel each day as you dragged slowly toward the weekend. When those guest speakers would come down for show and tell, they never ever in a million years told you how one day someone would pay you to sit still like a statue.

Sitting *is* the new smoking.

Anyone who has ever had a stiff neck, sore back, pulled muscles, frozen shoulder, or lower back pains from working in an office knows the curse of the computer. Regularly scheduled trips to the chiropractor, orthopedic surgeon, neurologist, physical therapist, and acupuncturist are evidence to the hazards of computing. Being younger does mean you have greater physical resilience but this environment is going to be yours for a few decades at least. Unless you decide to become an Amish farmer, you might start to have a large bottle of Tylenol or Ibuprofen handy in your desk drawer, purse, and/or computer bag.

An alternative to a career shift is learning to take care of your back health regardless if your employer has a program in place or not. To counteract the frozen chair syndrome, the most natural remedy is to practice stretching. Take a break every few hours and stretch for a few minutes even if it looks really silly. It does not take much time or floor space to do this but it will loosen you up by improving your circulation at the same time. As a benefit to you and the business, it will increase your productivity by enhancing your longevity and your mood as well.

This massive legal disclaimer is hereby declared for

those litigious loving individuals. Please consult your doctor before doing any exercises that may affect or aggravate your current well-being. Be sure to contact your manager and boss whether it is okay to take the aforementioned brief stretching breaks. If you share a small area with other coworkers, be aware of their personal space when you elect to perform any non specifically work related movements. You will also be personally responsible for your own actions, consequences, and healing while performing any kind of activity on, near, or away from the computer. Always enlist the aid of a trained professional who can answer all of your questions and predict those which you are about to ask.

With that out of the way, I can share with you what I do to take care of myself. This is based on frequently spending fifteen hours a day in front of my computer or someone else's for over twenty five years of my life. My personal experience covers training in yoga, pilates, weight lifting, boxing, and cycling. From extensive reading and a few thousand conversations over the years with personal trainers, marital artists, and massage therapists, I have been able to satisfy my obsession with body mechanics. I have learned my limits, how to extend them, and the repercussions.

To alleviate neck strain I do neck rolls and tilts to dynamically strengthen both the muscles and tendons. Gently tilting my head all the way back and all the way forward is the first thing I do. When I lean my head to the right, I massage my left shoulder at the same time. When I lean my head to the left, I automatically massage my right shoulder. Slowly rolling my head clockwise for a couple rotations lets the cricks in my neck loosen. Of course, I do it counterclockwise too for good measure.

For a good sitting spinal stretch, I keep my legs facing forward and turn my shoulders to the right so I can grab the right side of the chair back with my left hand. Ahhh, I can feel my vertebrae crackle. I face forward again and reflexively do a side to side neck tilt again for good measure. Now I turn my shoulders to the left and grab the left side of the chair back with my right hand for the opposite stretch. I come back to center and delight in the fact that you might be doing this just I am as I write this.

For my next trick I step away from my computer and

reach down to touch my toes without bending my knees. With each exhale I try to extend a little bit farther. Nowadays I can just about touch the tops of my feet with my palms. Before I started this as a regular body check, my finger tips were pretty far from the tops of my toes.

Tonight before I take a shower I may also do a back bridge for up to a minute. A bridge starts off with you on your back as you make a reverse table shape with your hands and feet. Yes, it sounds impossible to you but hopefully you will get over yourself soon enough. This feels great and has spared me from so much stiffness that I love this religiously. Before you sic a dozen lawyers on me, I will reinforce and repeat my earlier disclaimer by telling you to clear this with your doctor and physical therapist before you even think about this.

Back when I was working in a cubicle for a living I took it upon myself to sit on an exercise ball instead of a chair. Doing so forced me to constantly stabilize myself with my legs so I was never "slumping in my chair" like my coworkers. Of course I did not have a back rest but having to consciously remind myself to use good posture was partially self correcting. When I slouched forward or leaned back I would start to roll and I would have to make myself come back to center. The ball was also great for doing my stretching because it was supportive for my spine. I had a few followers who switched their chairs out and adopted my example too. As a free publicity plug, I strongly recommend the ones made by Theragear(formerly known as Sissel) because of their durability and high puncture resistance. High five!

Okay, now raise your hand if you have carpal tunnel, the most famous of the repetitive motion ailments diagnosed for the 21st century, or as the professionals call it, repetitive strain injury(RSI). This is when your hands hurt from typing or grabbing things all the time. Your poor widdle hands are prone in a latent claw position and the urge to cackle like a wicked witch is the last thing on your mind. Thinking back to the counter stretch theory, I have been able to hold that condition at bay by rolling my hands. I start by kneeling on the ground and holding my arms to my sides with the backs of my hands facing forward and my finger tips touching the ground. That is when

I slowly roll the backs of my hands onto the floor making sure my fingers are pointed behind me. Afterward I lift them off the ground and give them a good shake right at the wrists. The second stretch starts off with your finger tips on the ground again but this time your palms are facing forward. Flatten your fingers followed by your palm and you will also feel a slight stretch in your forearm. Shake it off again just like before. This is a preventative thing that I do as well as a treatment I taught my seventy year old mom for when her hands started hurting from gardening.

Big butt syndrome or the maximizing of the gluteus is another distinct hazard of being cooped inside in front of the computer. After spending all friggin day getting your soul sucked into that glowing screen, it is no wonder the gym is a foreign concept. Our hunter gatherer ancestors would be amused at the sedentary lifestyle people endure in today's quest for survival. Adapting to confinement translates to less overall movement and therefore fewer muscles being used, hence fewer calories burned; throughout the day you can also do some NEAT(Non-Exercise Activity Thermogenesis) or fidgeting for a calorie burn. The more restless you are, the more you will be able to fight the tendency to sink into your office chair. It may require hiring a personal trainer to force you to exercise once you have escaped the office for the day; one perspective is to learn good workout habits while another pragmatic view would be for you to realize the value of the money spent is cheaper than spending it on more prescriptions in your old age.

To fight the morning haze and afternoon doldrums, caffeine is a popular poison for many a cube warrior. Coke, Pepsi, Dr. Pepper, and the wondrous power of Mountain Dew fuels coders and desk jockeys all over the world. Starbucks has over 13,000 locations across the galaxy from where you can purchase their java bean beverages. No matter which caffeine laden sugary drink you dose yourself up with, the nutritional tax it provides your body is cumulative. The long term effects can be read in the gut of your nearby coworker who has been sitting at the computer a few more years than you have. "Oh, but I am drinking diet..." is such an incredible farce that I seriously look around for the ghost of Allen

Funt to tell me I am on a revival of Candid Camera. It is precisely because I do not have a PhD in biochemistry that consuming unknown chemicals is something I choose not to do. Industrial food additives have existed for only the last hundred years or so while the sanctity of a traditional food supply has sustained us for thousands of years. So for the sake of convenience, I suppose I could drop a buck in the soda machine and get a cold drink because it is right there down the hallway. I would also have to forget everything I have read about the possible carcinogenic and neurological risks associated with strange substances I know nothing about(go search the title word excitotoxin and the name Dr. Russell Blaylock and see if what he says makes sense to you). If you want to drink something with zero calories, try water.

When I am sitting in front of the computer, I also find the compulsion to snack a god given right. Maybe all those years of eating popcorn at the movies or munching on chips while watching television has creeped into my work habits. The salty goodness does seem to make me think better or it could be just a Pavlovian thing. More likely the intensity that I am using to focus on the screen causes me to ignore the subtle signs of hunger. Knowing that I am not ever breaking a sweat is a typical rationale for postponing a meal until I am really ravenous. Barbecue flavored kettle chips fulfill my desire for a tasty and noisy interlude amidst hours of diligent monitor duty. Strangely enough when I remember to bring pieces of fresh fruit and nuts to eat instead, I do not have that logy feeling at the end of the day.

Our requirement for sleep is something not shared by our super intelligent know it all artificial brain box. Your personal computer does not need to rest because of fatigue and can effectively operate continuously unlike you or I. Shutting down your pc is only needed when you want to reset the operating system or to save electricity. Puny humans absolutely need to regenerate on a daily basis for approximately eight hours at a time in a virtual coma mode. Any disruptions in a person's dormant cycle produces erratic behavior in the form of irritability, impaired judgment, slower reflexes, and decreased cognitive skills.

***

Let us review boys and girls!

Essentially how well you operate a computer depends a lot on how well you take care of yourself. It would not matter if you used a computer once and then abandoned it like the abs exerciser you bought from that late night infomercial; by the way, it is still sitting in your closet waiting for you to take it to Goodwill. However, this revolutionary invention though is saturating our lives and its usage is evolving faster with each passing week. Unless some catastrophe occurs on a biblical level, computers are becoming an intrinsic part of civilization as common as the toilet and those rolls of paper we use to wipe our derrière.

Understanding the basic architecture of a computer is as important as knowing your own basic anatomy. Avoiding poison ivy is at least as valuable as preventing spyware from taking over your laptop. Fixing a nose bleed is a basic skill akin to knowing what to do when your keyboard locks up. Adding a hard drive is much like minor surgery without all that blood and the necessity for an expensive anesthesiologist. Just as you can have lavish dragon tattoos, multiple piercings, and stylish clothes, you can paint designs on the outside of your laptop, customize your desktop with pictures of your children, and carry a designer computer bag.

All these parallels show us that what seems complex is within our understanding as long as we place it in the proper framework. We do not need to be a medical doctor or computer technician to maintain our bodies or our computers. Of course it is interesting to know how your DNA comprises your genetic makeup but in the course of your daily life, it is most likely an intellectual luxury. Same thing with computer code.

Know your computer, know thyself.

Operating the System

2

An operating system(OS) is the culture, the language, and the interface between the machine universe and you. It is the software that allows you to talk to your grandmother robot without having to know its primary vocabulary of 0s and 1s. Without it, you will find yourself staring at a blank screen waiting forever for something to happen. In a perfect world, every operating system should be so intuitive that a five year old can use it effortlessly and a drunken orangutan could fix anything that goes wrong. In case you have not noticed yet, we live on the other planet.

Any device fancier than your toaster oven which allows you to give it instructions beyond a handful of simple functions has an OS. So far microwave ovens, washing machines, watches, alarm clocks, pocket calculators, and coffeemakers are electronic items do not utilize an operating system. Desktop/laptop computers, point of sales systems in supermarkets that scan your groceries, personal digital assistants including smart cell phones, portable GPS(global positioning system) navigators, and modern automobiles all have some kind of OS. As an operating system's complexity approaches infinity, you will have something like the human mind; the movie Bicentennial Man starring Robin Williams actually shows what that would be like as a robot evolves into a person.

Typically we accept the OS on a computer as part of a package deal, like the way we accept a person at face value when we first meet him or her. Unspoken ground rules and personal expectations abound when we meet someone on a blind date. We expect someone to have good hygiene, to have a sense of manners, and hopefully be well spoken before we commit to seeing him/her again. Similarly we want a computer to be easy to use, to do things quickly, and to be simple to understand so we actually look forward to using it again. Suppose things do not work out between the two of you as the cliché goes. A simple breakup or maybe even a painful divorce separates you from your dates, acquaintances, and partners eventually. Unlike most of your codependent personal relationships however, your stoic computer has all of your finances, address books, photos, and TV shows under its control so you will always find yourself crawling back to it like an addict.

To make machines as pleasant and pain free as possible, programmers come up with improvements to the operating system that are both revolutionary and evolutionary. Radically new versions arrive every few years like changes in automobile body styles, even showcasing a major difference in the way the graphics look. Small patches to the system every month or so resemble adjustments to a ship sailing at sea, making sure its course is smooth. The faster the physical hardware changes, the more likely you will see modifications in the operating system. Translation: more complex machines need better brains to control them.

Here is the bigger pill to swallow for the non-technical boys and girls reading along with us on this not-so-yellow brick road. There are dozens of operating systems out there in the big bad world and there is no way you are going to master all of them for every single one of your little toys, movie players, musical jukeboxes, and portable workstations. Balancing the traditional needs of daily life with this genuinely artificial world is a losing battle which often relegates the computer to the "when I have time" corner. It collects dust and ages quickly in a case of technological progeria, while its resources lay dormant and untapped.

Patient parents have tackled the notorious manual and quick user guides with varied success. They find themselves skimming it briefly for the first time on Christmas day and denying its existence as the months pass. Achieving a working video camera in less than an hour is a battle won for a father and his brood. Getting to the special functions via a submenu is a daunting enough feat to put off until another day. Silent prayers to ward off the specter of an unknown error are practically universal. When the first critical error message is displayed, fidgeting with the controls at random until it goes away is really what everyone does.

Nowadays your new computer does not come with a operating manual per se. To save on printing costs, they encourage you to go online for support or to consult the built in help software. No manufacturer is going to include an actual paper manual that tells about your computer in depth because it would also be too huge. On my bookshelf I have an outdated "tips and tricks" book from 1995 which is bigger than the New York City Yellow Pages. Your local chain bookstore has a wall of computer books that barely touches the surface of what computers can do now.

Pretend you suddenly dropped someone in the middle of Paris who has never read or spoken a word of French in his life. The first few minutes are disorienting because you need to find a restroom but the signs are all in French. If you are lucky you find yourself next to a McDonald's where the symbol on the doors tell you which is the men's room. You can also order some food because the menu is straightforward if you take the time to pick something recognizable. Unfortunately you have no cash on you but your credit cards are still working even if it takes you a few minutes to understand the instructions on the keypad to proceed. A stroll outside leads to an open market down one street with a sign pointing to the Eiffel Tower in the other direction. As you look up and wander toward one of the most recognizable landmarks in the world, a thousand questions come to mind. How do I hail a cab? Where is the Louvre? Is it going to rain later? Is this street safe? Does that hotel have someone who can speak English?

Poof.

Suddenly you are sitting in front of a computer for the first time. After you turn it on, you are a little disoriented by all the colorful symbols and boxes that pop up. Some of the instructions make sense while others seem like they are in a foreign language. There is an icon on the screen interspersed with the word "Internet" which flickers like a tiny neon billboard. Touching it causes the entire screen to fill up with a rectangular window showing a directory of topics ranging from the weather to stocks to movies to recipes of the day. On one side of the page there are advertisements for children's chewable aspirin and a small video showing today's news headlines on the other. As you start touching phrases on the screen, other windows start popping up accompanied by a the sound of Vivaldi's Four Seasons overlapping an announcer telling how you have just won $50 in gasoline if you just click here. As you try to makes sense of the cacophony, a thousand questions come to mind.

How do I turn the sound down? What do I do to close all these messages? Can I really get $50 worth of free gasoline and what is the catch? What company did they say had lead paint on their toys? How does Google Earth create such a clear picture of the Eiffel Tower?

Immersion is the best way to learn in either case. It still takes a number of years to absorb and breathe in all the facets, details, and nuances. To become an expert requires years of study of the standard language, picking up street lingo in context, and striking up conversations with aficionados who relish the slang. Knowing the history, including key events and recognizing famous figures in the field, also helps to fill in the picture. Tell yourself it is okay to just let yourself be an open minded computer tourist.

As your verbal bartender and technical tour guide, I will serve up aspects of the most popular operating systems out there. Like sampling drinks at your favorite tavern, boba bar, or coffee house, you have to taste enough to be able to describe it to your friends. This will not make you an expert at techspeak but it will provide you with some of the basic concepts in a language closer to home. With the power of a little creative wordplay, dry textbook facts will be livened up with some juicy metaphors.

Computers became personal when they got sticky and gooey. By gooey, I am referring to a pronunciation of GUI, a Graphical User Interface, a screen with pictures instead of just words. Repetitive mindless work is more tolerable when you can sneak in a game of Solitaire or two. A modern word processor has built in white out, a dozen highlighters, a thesaurus, a dictionary, and an infinite number of typesets whereas the first few generations were not much more than a typewriter. The stickiness factor has to do with when addictive applications started stealing time away from watching television. Email, chat rooms, and real time games with other people created one generation of addicts. Browsing through other people's psyche through blogs, shrines, and web rings detailing rich fantasy lives captured another segment of the population. Any movie you can think of, any song ever recorded, and any television episode ever produced for free and whenever you want makes a computer a valued family member.

Once upon a time when computers inhabited whole rooms and chewed on punch cards for data, they were solely the domain of die hard scientists and engineers. Just prior to the introduction of a GUI, talking to a computer required reading and writing lines of commands on a monitor to do everything. Thanks to thousands of programmers all over the world we get to navigate though pictures instead of having to learn the language of code. For the majority of the population, we will never have to see what is behind the Wizard's curtain to enjoy Oz.

When the power is off, a personal computer is as unresponsive as a body in the morgue. Turning the power on wakes the machine up from its dead state, initializing the BIOS (Basic Input/Output System)sequence. The BIOS checks the status of the basic hardware components that are connected to the motherboard one by one. Those lines which go speeding by as soon as you hit the on button give you details about the size of your memory, processor type, and types of drives attached. Advanced technical people have a two second window to hit a certain key to access the menu that lets you monkey around with the settings. Standard practice is to ignore it while you wait for the colorful people friendly OS to come up.

An off the shelf or prebuilt personal computer will immediately begin loading its operating system after the BIOS finishes. Here is where you get to see the big splash screen advertising the name of the OS that you use every single day. To access the network, a multiuser environment, or the default secure setup you need to enter(or select) your user name and type in your password. For those happy go lucky users who have the login set up as automatic, you get to go directly to the desktop!

Let there be Windows

Microsoft Corporation produces the most common operating system in the world. Its computing OS called Windows has a number of versions through the years including 3.1, 95, 98, 2000, ME, XP, Vista, and Win 7. It also makes an OS for servers (computers that manage networks) and portable devices like PDAs. Assign a placeholder cost of $100 per computer and you have a conservative picture of how its founder is the richest man in the world.

Microsoft Windows is characterized by a Start button on the bottom left hand corner of the screen. Throughout the last few versions of Windows it has changed shape but remains a constant fixture with its logo displaying greater prominence; at the core of its omnipresent logo is a red, green, blue, and yellow flag that is supposed to look like a flag waving in the wind. As the starting point of the system, it allows you to access your programs, settings, files, and ironically the command to shut down the computer.

Assuming a touch screen or tactile holographic display is not available, you will be using a mouse(or touchpad or some other movement based input device) to control a pointer. The pointer will generally look like an arrow, a hand, or a large "I" depending on where it is on the screen. That arrow is the one you see the most because it is the default symbol that enables you to access everything. When you are on a web page it turns into a white hand as you hover over a link, signaling that you can click on it. That "I" means you can activate the cursor so you can start typing to your heart's content; once upon a time the cursor was either a blinking solid rectangular or a mere blinking underscore but nowadays it is a vertical line that blinks or it is a vertical line that just sits there... oh, and it can also be invisible.

Like having an illustrated restaurant menu in a language you barely speak, you can select what you want by pointing at it. Clicking on the program, a file, or shortcut you want tells the invisible waiter that you want a painting program opened up or a word processor started. A tiny icon next to a brief description is the kindergarten equivalent of learning basics shapes or colors. By hovering over any of the pictures or symbols for a full second you might see a longer description and/or the path to its location, whether it is on your hard drive or on another computer elsewhere.

On a typical personal computer there are literally thousands of files that reside on the hard drive, representing every single image, document, program, song, and video available for you to access. Out of sheer necessity someone decided long ago that it would not only be helpful but absolutely necessary to create some kind of file system to efficiently find what you wanted. No one wants to look through anything longer than a grocery list to find something, especially not a dull computer screen. For those fortunate enough to have studied biology, it is a cool side note to pay homage to Swedish scientist Carl Linnaeus for his detailed filing system which classifies all living things.

Suppose you walked into a large room with papers scattered across the floor like autumn leaves fallen from a gathering of oak trees and you needed to sort them. After you got over the anxiety of how you arrived, hopefully you would be grouping the papers in large stacks according to category. Feeling a little more ambitious you would separate those into smaller more specific piles which could fit into manila folders. The floor is visible and now everything fits nicely in a holy filing cabinet like the one God keeps in the movie Bruce Almighty. Your computer's files are neatly sorted for you in much the same way, nested folders virtually residing on your hard drive.

Oh sure, you can use the search function to have the computer find the file you are looking for by partial name, date modified, size range, type of file, and even general location. As an everyday practice though, you do not ask the librarian for help to find a book every single time you go to the library do you? Just remember that more frequent trips to your favorite places will make you more familiar with their location. This works with driving to new neighborhoods as well as locating the inspirational songs on your personal computer.

The major benefit to having your own PC is that you can organize your files and folders by doing a "drag and drop" or a "copy and paste." They are not dance steps or code words for entering a clandestine scrapbooking club either. Borrowing the theme of a tiny little nineteenth century book called Flatland, we have to shift our thinking from a three dimensional world to a two dimensional one. On a real desk you could grab the electric bill, a pencil, and stray vacation picture from the corner of the desk and put it in a manila folder without a second thought. On the computer, you are using a mouse to select a group of pictures by highlighting them one by one or in a group and dragging it to a folder while holding the button and letting go to drop it. This virtual "drag and drop" method is analogous to what you are doing with your hands when you physically move something on your desk. Unlike the real world, we can create exact copies of things and destroy them with ease on the computer. Hence the "copy and paste" function is like me touching a glazed donut for a moment and pointing to a my plate and making another one appear by magic.

Based on the rules of the computer world you can also create folders, blank documents, shortcuts, and plain text files right out of thin air. They can be born on the desktop or inside a folder with a simple right click of the mouse. A right click on the newly made icon and it can be named faster than your baby at the hospital. Your files, folders, and shortcuts can condensed, consolidated, and deleted outright by a sweep of your wrist and some clicking. If cleaning up your closet was half as easy as this then maybe procrastinator would not be your middle name.

For the truly efficient minded or perhaps just anal retentive, the use of shortcut keys is an indispensable time saving practice. For the majority of us, knowing just the most common ones are enough to make us happy campers. Hitting the right combination of keys to elicit a function on the keyboard is reminiscent to finding those special moves in those arcade games from the 1980's. There are dozens of sets of keystrokes that work in Windows as well as across applications. Many pieces of software also have their own series of shortcuts which are unique to that program. In life we probably use gestures in the same way to convey something quickly to one another. Similarly, some gestures are universal and others belong to single gang, tribe, or culture.

Going "mouseless" is an advanced test for people who know their keyboard shortcuts well. Should your touchpad, trackball, or mouse fail on you, it becomes helpful to know how to navigate through your computer using only your keyboard. While the grandparents complain about walking five miles to school uphill both ways, the next generation gets to moan about memorizing key commands and keyboard shortcuts. Kids today do not know how easy they have it with their pointing device and graphical user interface. Although the savvy user will take the time to learn things old school.

Back in the day, there were also a lot fewer types of files

to contend with. Not only have the number of different kinds of programs for the good old PC expanded to cover every function imaginable but the variations of a single kind of file have increased dramatically. A few decades ago when you saw a .wav file you knew that it was a sound file because that was the dominant audio type. Nowadays .mp3 files are so common that that have leaked into the English language as simply MP3s. Other common types of files are .flac, .wma, .ra, and .m4p followed by a slew of lessor known ones including .aiff, .au, .vox, and .gsm. Due to the rise of digital cameras, the popularity of the .jpg file has become a given, almost completely overshadowing the others except to those in the graphic arts community. Even with Microsoft's practical monopoly on the word processor with its Office suite, Microsoft Word still has a quite a number of file types you can save your work under.

Unfortunately for the casual computer user, figuring out the differences between one type of file and another can be as tiresome as identifying mold spores under a microscope. As each file type has specific attributes to consider, knowing which one is best for your needs requires doing your homework. Thankfully the informative nature of the internet also takes the guesswork out of explaining the nuances between a .flac file and an MP3.

The file extension is usually a three letter acronym or shortened version of the actual file type. As you can see from the previous paragraph, the suffix can really be any alphanumeric sequence so the rule is not set in stone. Conveniently the operating system will automatically detect what category a file falls under based on its notation and associate it with the right program. Unknown file types are left alone and the PC will prompt you to select a program if you try to open it. When you are saving your work in a program, the software will use the default file type most commonly associated with it. At that point, you can choose to save it as a stripped down version which uses less space or as a more complex file to preserve the original level of detail.

Another symptom of the personal computer's increased complexity is evident in how it tries to talk to every device that you attach to it. Bragged about as a having "plug and play" functionality, the computer's operating system should be able to recognize whatever you attach to it and immediately start to work. To make this happen, the OS has a massive list of drivers which corresponds to every single device(it is pronounced pe-riph-er-al) approved by Microsoft as a compatible partner. The driver is the PC's own instruction manual for the printer, flash drive, web cam, and whatever device you connect to it.

Keep in mind that the programmers who write these drivers are fallible too so the one you have may just be an early draft. Usually the version that reaches the public is final enough to work perfectly fine. In the event that your device seems to be having a hard time communicating with the computer, it may be time to look for a updated driver from the manufacturer. Reputable companies provide all the latest drivers for their products on their web sites so you do not have to fret if you have lost the CD that came with it in the box.

Switching to the software only side of the world brings to mind a five minute lesson on codecs. My hunch is that you do not care that it stands for COmpressor-DECompressor or that it is a piece of software used to decode a audio/video file. When the video player on your computer is missing the sound and/or the picture and you get some message about a codec thingy, you will wish you could remember this paragraph though. Admit to yourself you are a video junkie and the real reason everyone watches less television these days is that people are watching stuff on the computer. Where did that accusation come from you ask? Your standard player will play the ordinary files on the DVD you bought from the store without a hitch. Watching a movie you downloaded on the sly or got from a friend may lead you to seeing a message about a missing codec. Either you can install an open source video player with all the newest codecs built in or you can blindly look for the exact codec you are missing; sadly the player does not know otherwise it would have gotten it automatically for you. Today my favorite universal video playback software of choice is the very versatile VLC player.

To say that everything is technically a work in progress and therefore requires periodic updates is a fair statement. The operating system needs patches to remedy the holes in its security where a devious hacker might break in. Your anti-virus software needs weekly if not daily updates of all the newly discovered virus patterns so it can protect you from the bogeyman. Sometimes your router, cell phone, or PDA may even need a firmware upgrade to solve that weird glitch that never seemed to go away.

Seeing how you can tell a computer to do things for you on a schedule, making it perform automatic updates seems like a no-brainer. However having the maid clean the floor while you are cooking dinner is kind of chaotic and forces both of you to take twice as long. Even if the newest computers brag about being able to multitask well, you will notice the system slowing down when you try to do your normal work and perform software updates on your machine. If your computer is older and perhaps a bit error prone, diverting system resources to do maintenance while you are driving at freeway speeds may just cause you to crash. How many times have you seen other drivers put on makeup during your morning commute?

Setting aside time to brush your teeth on a daily basis and seeing your dentist periodically are deliberate habits we learn to maintain our dental health. Typically we will set up an appointment for a regular cleaning by a trained professional to take care of things we are not equipped to deal with. The same scenario can be transposed to taking care of your computer's health. Spend a few minutes a week to check for OS and program updates when you are not checking your email, paying your bills online, or working on the family photo album. Taking your computer for regular checkups may seem like an added expense but to some people, losing all their data can be more painful than getting a root canal.

Unless you have a decent background involving TCP/IP protocols, contracting a professional for your networking needs would be a wise investment as well. Sharing files through your router at home is an easy do-it-yourself project for the normal user. Creating multiple user accounts on your home machine for other family members to   have their own desktop setups is another one. Configuring a small office for a few machines to access a server or troubleshooting how to run an application at your house remotely from a your hotel room are worth calling for help.

Helping computers communicate with each other will

always require knowing their native language. At some point in our life, we are all wide eyed children gazing at the computer, as some towering adult helps us with our homework, entertains us, and connects us with the world. Learning about the networking aspect brings us to an instant grown up status because the concepts are more complex and you get to use big words.

MAC Snack

Apple makes the prettiest computer in town. On the outside, virtually everything they make has the feel of a piece of modern art. When it is turned off, it rests on your desk literally like a manifestation of beauty, designed to accentuate the room like Rodin's plaything. During use, its desktop interface is rich with vibrant color and even the icons themselves have a three dimensional shiny new toy quality.

Buying a Mac is also an expensive purchase. Compared to a Microsoft Windows based machine, you will generally be paying about $400 or more for a similar hardware configuration. That figure obviously has some wiggle room but the pricing for an Apple computer is consistent across the resellers and at their store. In contrast, Microsoft machines are sold at various prices depending on the manufacturer, retailer, specific model, sales promotions, and time from initial release. There are no compatible clones out in the world for the consumer to buy either so the concept of Mac against Mac competition is moot.

Marketed to the masses as the hip cool technology, Apple's portable music player and cellular phone have made them more than just a computer company. Their iPod has an attractive aura surrounding it even though there are other MP3 players on the market with more features and half the price. People were waiting in line overnight in obsessive anticipation for their iPhone which was marked up by several hundreds of dollars according to several teardown reports; a teardown is a dissection of a product for analysis of the individual components, cost, and assembly usually performed by someone other than the manufacturer. Another offshoot to their core business is their successful iTunes online store which sells digital songs and TV shows, proving to the business world that consumers are willing to purchase downloaded music en masse.

When it comes down to the computer itself, the watchword is a constant mantra of elegant simplicity. There is one application to listen to music, one to watch videos, and one to manage your photographs. If your needs are met then it is the best computer in the world. Should you want more software choices you should keep in mind that not every software company makes a Mac version. There are a couple programs out there that lets your Mac pretend it is a Windows environment but they do not always run as cleanly as if you had Windows for real.

Inside the modern Apple computer you will find all the same kind of innards as in a PC but they are for the most part not interchangeable. A motherboard, RAM modules, a video card, and optical drive fit in there quite well but the home user is not welcome to tinker. Any upgrades not performed by the Apple people are frowned upon and might void your warranty for all eternity.

Mac users themselves comprise less than ten percent of the active computer user population. Interestingly enough they have banded together to support each other in user groups since the dawn of the personal computer. As a decentralized cult and fan club, members are freely encouraged to join together by their almighty corporate god. They have been known to trade tips, stories, and freshly baked cookies at many of their local chapter meetings. Although my experiences have been strictly PC, I will be the first to admit to having experimented with Macs.

Through a brilliant combination of philanthropy and marketing savvy, Apple has made many donations over the years to public schools across the United States. Their generosity has introduced an untold number of children and their impressionable parents to the world of computing. To the company's immense benefit, this embedded the Apple brand onto millions of young consumers at an early age plus it encouraged further sales to schools wishing to expand their infant computer labs. The Pied Piper had made its way into the long term strategic vision of the Silicon Valley boardroom using a shiny Apple.

As for the Mac's operating system, it evolved from version 1.0 through the next set of revisions with a whole number change for every major update. When it got to ten, Apple started calling it OS X and assigning it a feline code name. In the last few years, large improvements have been notated by a 0.1 increment and a different kitty cat. As of 2012 they are up to Mac OS X version 10.8 advertised everywhere as Mountain Lion. So far they have used Cheetah, Puma, Jaguar, Panther, Tiger, Leopard, Snow Leopard, and Lion. When they decide to release Calico, I am going to get a Macbook.

The hunt for the power button might be tricky the first time because it is situated with a clean design in mind and not gaudy obviousness. After you hit it and the Mac finishes booting up, you will notice a row of large colorful icons on the bottom of the screen. This dock holds shortcuts to the most popular programs you are likely to use and will appear magnified when you hover over one. On the top of the screen you will find a thin gray bar with the time, day, and status icons on the right side. Appearing on the left side of the top bar is an apple symbol, Finder, File, Edit, View, Go, Window, and Help. From those panels you can change your settings, find your files and programs, manage your work, and look up help on how to use the computer.

Another striking difference you will notice right away is the cyclops mouse. For anyone used to the two button mouse with Windows, becoming familiar with the single button action of the standard mouse is a sticky treat; you can mimic the special function by hitting CTRL as you click. At least that was the case until Mac's Mighty Mouse came along in summer 2005, a two button wonder sometimes sold separately. Coincidentally, the tiny rollerball on top does make it kind of resemble a lonely eye.

Much of the rest of the general navigation of the operating system is similar to the other OS made by those Microsoft folks. Drag and drop is a common exercise to both systems as are nearly identical shortcut keys. Scroll bars to move data across the screen, built in screen savers, versatile fonts, using a photo as wallpaper, support for dozens of languages, and playback of DVD movies are also shared elements. Although Apple is susceptible to viruses too, there tend to be much fewer targeting them. As for which OS crashes more, picking a side on that debate will earn me more bruises than rooting in an English football match.

Unix, Linux, and DOS mix

Let us call them operating systems on the fringes of the universe. For the average citizen of the world, you are not going to care about Unix much. Its cousin Linux is getting more frequent mention in the news though as a replacement for Windows. At least you should know that Linux is not that kid in the comic strip who is always carrying a security blanket around and getting crushed on by a little girl named Sally. Poor old DOS lays quietly underneath your Microsoft Windows OS like a dusty book on the shelf. Unless you are a historian or technophile, the other few hundred are barely up on the mass radar.

As the mainstay of workstation(big...for the lone techie), server(bigger...to serve many techies) and mainframe(biggest) computers, UNIX was developed in the late 1960's as one of the more significant operating systems. From the original project, there have spawned several versions over the years by various companies, which expanded in features as computers themselves evolved. As these were developed privately, the instruction code was considered proprietary to the business who developed it. The source code as it is called was considered closed off to the public.

In the 1980's a group of engineers decided to create the Gnu project which would develop a free operating system where anyone could view its code, modify it, and publish it without obligation. As a revolutionary step towards collaborative social technology, a manifesto written by Richard Stallman was published which advocates against the commercialization of computer software. Open source software by definition follows this doctrine and has produced Linux as well as other offspring; we will come back to this gem later I swear.

The openly customizable nature of Linux makes it attractive to hobbyists, gadget manufacturers, and businesses alike. For the electronics company, an open source operating systems means they can modify a free template for their specific needs in order to cut costs; this savings often is passed down to the consumer who would pay less for a computer, phone, or other device with a Linux based OS. As a learning tool, it is foreseeable that modern teenage students would play around with their own version for a class science project. Instead of gathering around a garage to tweak the carburetor, nowadays car guys can connect to the car's onboard computer to change the parameters of the computer controlled fuel injection system. Homebrew rookies are discovering that a Linux installed personal computer is safer, cheaper, and most of all faster than one running either a Mac OS or Microsoft Windows.

As a recap reminder, an operating system is just another piece of software, in its most basic form, it is just a list of instructions. At the fundamental level, it is the primary set of directions for the computer's cpu brain which lets it talk to its eyes, ears, and hands. Generally it also tells the brain to coordinate the rest of the body and do chores, aka applications. The OS lets the robot walk, sing, and chew gum at the same time.

DOS is a simpler operating system which shares the primitive characteristic of a barely trained pet. You can tell it to fetch a stick, sit, or speak but you are typing out each command one at a time. Having poor Fido, excuse me DOS, do a combination of tricks or tasks is really asking too much. When you need something easy like a hard drive formatted, your DOS is a reliable companion. Appropriately enough, single word commands are exactly what it was taught at the obedience school too.

Puter Linguistics

Whales communicate underwater by use of song, which we record and playback for mediation seminars. Bees move around in a dance like motion to show each other where food is located. At some level, all creatures convey messages by gesture, facial expression, and speed of movement. Humans over the course of thousands of years have developed countless spoken languages, dialects, and just plain emotive sounds. Perhaps a few hundred written languages are in active use today, with the roots of some dead ones evident across several cultures.

Machines have evolved from mechanical clocks and tractors to electrically controlled creations with a brain relative to our own. Skipping a soul debate, we can acknowledge the computer's ability to take instruction is predicated by its usage of language. On a transistor level or at the computer stage, a set of electrical pulses both yield an expected reply back. The intermediate communication between the machine world and ours is naturally a programming language.

In the short span of decades, the number of computer languages already surpasses a few thousand. Of course, the average programmer will actively use only a few even though he/she will encounter many throughout his or her career. Typically a program is written in what looks like a pseudo English shorthand and translated into a code that is pretty much unintelligible. Each computer brain model understands a particular machine code almost like the way each generation has its own set of slang. Adept programmers can recognize the difference in syntax as much as a Portuguese speaker can identify certain aspects of Spanish.

Schools with a foreign language requirement would do well to encourage learning a computer language such as C++ or SQL to strengthen a student's background. As machines in general become more advanced, interacting with a number of electronic operating systems will be an inevitable aspect of our lives. An understanding of the computer world will be as important as communicating with our neighbors across the street, over the ocean, and on the other side of the Earth.

The Library Infinite
3

"I just need something to check email and surf the internet," is the line that echoes across the mountains and through the valleys. This is what you hear when someone goes shopping for a computer or when they ask someone to help them buy one. For the most part that is what people do with their computer now, aside from giving their cat a comfortable place to sit. People who use their personal computer for work will just run the one program that helps them earn their paycheck, exchange emails about when the next project is due, and surf the net.

If I have one piece of advice to you unrelated to sunscreen, it would be this: learn how to use a personal computer. Surfing the web is fun and occasionally informative but there are an infinite number of other things your computer can do too. Bear in mind that it should never be a replacement for actual human contact, real interaction, and tender relationships. Next to your swiss army knife, your personal computer is the greatest tool you will ever own. I will elaborate in detail now...alphabetically....

Arcade games from the 80's are available at your fingertips without having to feed the countless supply of quarters. Atari 2600 console games, Colecovision, Intellivision, and the Nintendo offspring are playable using a type of software called MAME(Multiple Arcade Machine Emulator or as I like to tell people, that program that lets you play those old school video games). They are designed to faithfully look and play exactly how they were made originally. So what if the graphics are primitive by today's standards, the gameplay is as addictive as ever.

Accounting and tax preparation software lets you obsess over your finances by being your own bookkeeper and speedily filing your taxes online. Paying your telephone bill, cable TV bill, electric bill, water bill, cell phone bill, Visa card bill, Discover card, AMEX bill, Mastercard bill, magazine subscriptions, medical bills, Costco bill, Macy's bill, Nordstrom's bill, Target card bill, car payments, car insurance, and various charity donations by computer saves you up to seven or eight glorious dollars in postage a month! Assuming you have the money available, you can schedule the funds to be paid the day it is due to hold onto your money as long as possible while avoiding late fees; paranoid fears about server errors are ranted in chapter 8 so cool your jets. Monthly budgets scrawled on scratch paper at the end of the month can be replaced by a real time snapshot of your debt at any moment of the day or night.

Right now you can order a custom fitted pair of denim blue jeans using a basic computerized scan of your body, read various biographies on blues legend B.B. King, or find a new powder blue bassinet for that beautiful baby boy. These pants specialists record your measurements and style preferences, transmit the data to seamstresses, ship the order directly to you, and keep everything in a database for repeat purchases. In a similar vein, a company in Japan makes automatic computer controlled machines which can knit seamless sweaters just like a Thneed made from a truffula tree. Shopping and shipping are two terms that go hand in hand with using a computer to get your baby gear. Going online at your messy apartment, neighbor's house, office cubicle, or local coffeehouse means you can shop from literally anywhere. No matter when a store is actually open for business to the public, their web site can take orders 24 hours a day despite that it may be in different time zone. As time progresses it is harder and harder to find a company without a web site, no matter what part of the world you find yourself. A Korean tourist traveling in New Zealand can use her laptop at the hotel to order a handmade crib from a carpenter in London to be shipped to a friend in Columbus, Ohio.

Back in the days when calculators were the indispensable tool of every rocket scientist and chess was the king of war game simulation, a computer far less powerful than your ~~Blackberry~~ iPhone would fill up a large room. Nowadays, every aspiring astronaut can access the built in calculator in any laptop, desktop, or mobile device. Those fancy scientific graphing calculators that display parabolas and other curvy lines from your algebra test equations are obsoleted by easily downloaded software for your home pc... Playing chess against the computer and winning at the advanced levels was the alpha geek's feather in his cap. Of course the advent of the internet allows you to play against people internationally at any time of day(that goes for checkers, backgammon, go, Monopoly, Scrabble, and every other conceivable board game every made).

Dictation machines no longer need someone to transcribe hours of notes with the magic of a stenographer in a box more commonly known as speech-to-text software. As the accuracy improves along with the inevitable increase in computing power, we will have robot servants in no time. Voice recognition is tricky due to the variance in human speech patterns, vocal quality, and the sheer number of distinct languages(not to mention dialects, slang, and pigeon blended versions).

Encyclopedias that once took up a long row or two on a public library bookshelf had a brief life on CD and DVD, compressing a few hundred pounds of books to a few shiny plastic discs. Although this tangible media can easily hold video interviews, archive documentaries, and complete discographies, the next step are portable devices that can access the infinite library called the internet. In the 1970's, comic book writer Jack Kirby described a race of people called the New Gods who carried unique portable supercomputers named Mother Boxes. We are there.

Exercise is the most ironic subject that we can possibly associate with the computer. Thousands of types of martial arts, yoga, weight lifting, and sports can be demonstrated by video, animated graphics, and pictorial diagrams on your humble notebook. Secondary devices can be connected to the computer to measure your body fat, weight resistance levels, carbon dioxide output, heart rate, bone density, height, mass, and calories expended. Countless books, articles, and advertisements on the fastest way to achieve a trim body, become a muscular hulk, or look ten years younger are all available at your fingertips online. Unfortunately, it is well documented that the higher usage of computers by number of hours per week is a significant contributor to us becoming collectively softer in the middle.

For the fabulous world of dairy foods, French pastries, and other fine edibles look no further. You can discover every scrumptious one of Ben and Jerry's outrageous flavors online or check out the history of the simple butter croissant. Food festivals are showcased by time of the year, geography, and gastronomic category. Each of the In-N-Out special order requests are listed on several web sites. Recipes from every country on the map, including vegan, kosher, and low salt versions are available too. The IFCE(International Federation of Competitive Eaters) has a home page that shows how you can prove your ability to gorge like a madman.

Gardening is a peaceful domestic activity, cultivated by the rise of suburbia and the urge to decorate your backyard like a domesticated jungle. Landscaping software will let you plan out every minute detail down to the petunia patch next to the koi garden to the organic vegetable trellis filled with baby squash. Hundred year old seed companies like Burpee, Park, and Henry Fields print out gorgeous catalogs that are now accessible by internet data delivery. For the dirt plot deficient, hydroponic kits for herbs and illicit hallucinogenic foliage alike can be found via the net. Topiary mazes, record breaking homegrown tomatoes, crab grass treatments, and delicately cultivated bonsai trees all reside on the web.

Hair care is a personal habit that powers a multi-million dollar industry, all of which is entangled by the computer's influence. From discussion forums that chat about wigs for cancer care patients to holistic ways to fight graying, everything has some kind of connection. Certain salons have imaging software that can show you what you will look like before you get that hairstyle; you can also download consumer versions to play with different looks at home. Not only does your shampoo bottle have the company web site on it but you can also get coupons by email when you sign up for their monthly electronic newsletter. Those stunning models in their advertisements have also had their picture retouched before it went to be printed. Behind the scenes, retailers are tracking that blow dryer in their inventory tracking system so they know whether one brand sells better than another. Your store membership card can also be used to send you recall notices on bad product because it records every single item you have ever purchased from them, including that nose hair trimmer that was on sale last year.

For people that love Indian or Italian culture, staying in

touch is even easier with the benefit of instant messenger software. An instant messenger lets you send text and share pictures at the same time in a staccato telegram style. As long as the other person is logged on the network you can trade quips faster than tossing a tennis ball. An icon, avatar, or other tiny picture tells you if your friend, colleague, or ex-boyfriend is around to chit chat. Yes, you do need to get their permission to add them to your list of friends first otherwise it is still considered a minor form of stalking... Regardless if you are in New York or Los Angeles, the Festival of San Genaro is still in September and a quick online search can tell you what part of the city it is held. Vacation plans to Milan or Rome are easily made, paid for, and checked online quickly by that friend who picks you up from the airport. Whenever I feel like having some gnocci, I can find a hundred variations on the recipe and any restaurant within a 50 mile radius of my zip code that serves it. Italy's history, local customs, weekly weather, and concert venues can be perused during the overseas flight using the embedded screen on the back of the chair in front of me.... Learning about the infinite dance numbers of India's film culture, flatteringly known as Bollywood in deference to you-know-what, is an adventure made easier with the internet as a tour guide. As a student of the world, the contrast of poverty and modern technology is all too visible through the world wide window. For any seasoned chef, the diverse range of curry spices is a boon to the palate yet it literally demands the resources of a computer to catalog.

A vast repository of jokes is secretly what the internet was originally intended for by a sinister society of jesters dating back to the fifth century. No, not really. That is an urban legend just like the one about waking up in a bathtub full of ice with your kidneys removed. Internet hoaxes replaced chain letters as the pastime of the gullible and truly idle. Scary campfire stories, blanket prayers, and quirky jokes will make it to your email inbox by many of your bored friends and relatives. Riddles, limericks, anecdotes, nursery rhymes, one liners, as well as kids' jokes, blonde jokes, gay jokes, dumb guy jokes, ethnic jokes, and "you know you are... if you..." lists continually make their rounds. As the ultimate time waster, you can spend hours at a time reading through numerous websites that archive this cultural sawdust. Of course I have read thousands of those jokes myself but it was purely for research purposes only.

Have you prepared your child for kindergarten yet? Five year olds are learning their alphabet from a keyboard rather than flashcards, their shapes from Elmo on the computer not Sesame Street on PBS, and the color blue from Microsoft Windows as opposed to the cloudless sky. In addition to the normal childhood warning about strangers, nowadays we have to teach kids that the bogeyman is out there as an online predator too. Narnia's wardrobe closet portal is passé compared to a typical desktop and monitor because the real world is what awaits on the other side.

Every public library in the world is connected together, creating an infinite collection of books, magazines, newspapers, CDs, DVDs, and tapes. Initiatives are in place to preserve everything digitally so that instead of checking out a physical book from across town or cross country, you can download a copy onto the computer in front of you. Whereas one book can change the way you live by teaching you a marketable skill, opening your life to a new point of view, or showing you how to choose healthy foods, several can reshape the world. A library infinite just may save it.

However most of us would rather be entertained by a massive movie jukebox so we can just forget daily life and fall into a fantasy for a few hours at a time. If this magic box lets us watch any TV show ever made, regardless if it has been broadcast or not, that is even better. As Homer Simpson wistfully said, "Television! Teacher, mother, secret lover." Our codependent relationship with the boob tube has been supplanted, like trading a morning coffee habit for a heroin addiction. On one hand, the convenience factor is amazing in that we can practically watch anything we feel like whenever we want via the internet genie. Abusing that luxury is transforming us into a full fledged race of couch potatoes, sucking down snacks to feed our slowing metabolism. When your eyeballs are locked into that frame of space and your breathing lowers, you have achieved a trance state. Sit off to the side and observe someone as they are watching TV or surfing the net and you will know exactly what I mean.

Magic mushrooms and music boxes litter the psyche of many an artistic soul and serve to enhance life's journey. Adding a fungus to our menu yields superb flavor, death, a psychedelic trip, or cancer fighting compounds, depending on whether you pick the right one. At the very least you will not impact your diet since these spore grown wonders provide absolutely no calories to your daily count. Aside from buying them a reputable market, it is helpful to educate yourself so that you do not end up in the hospital. Becoming an amateur mushroom hunter is something that you can do by finding others with more experience via the internet. Through the collective travels of others you can avoid the mistakes they made, learn the best places to find your fungi, and most of all how to tell the difference from the good, the bad, and the bland... Whether your music box is an old school Walkman, an iPod(or one of the dozens of other less famous competitors), a portable stereo(once called boom boxes because people often cranked up the bass so you could feel the shock wave from across the room...and supposedly guys carried them on their shoulder like a 2 by 4 too), or a wind up wooden antique from your grandmother's house, it will still cause memories to pour out like a faucet. The emotive force of music is felt by every culture throughout time from the moment someone stopped to listen to the rhythm of his own heart. We can dissect the pitch, tempo, and melody of an orchestra of instruments with an entry level computer. Add in a little expertise, a basic grasp of music theory, and some store bought software and you have a home studio that rivals the major record labels. Selling songs you have recorded, edited, and published on a universally accessible network means a talented musician can make a living without a record company to distribute physical records, tapes, or discs. A live band can be carried in your pocket because music has been translated into the language of data, which is easily stored, transmitted, and replicated. So far the collective sensations of a live concert performance is something yet to be mimicked.

Nurses provide an irreplaceable link between the medical establishment and the vast needs of a suffering patient. Advances in imaging technology have made x-rays and MRI diagnostics safer and cheaper. Innovative drug therapies are developed by the analysis of thousands of hours of clinical trials and laboratory tests. All of this information is shared by researchers, technicians, and doctors at the click of a mouse. Online journals, symposiums, and seminars keep your hospital administrators, doctors, and nurses versed in the latest treatments, drugs, and surgical procedures. Robotic assisted surgeries are even made possible by surgeons using remote controlled scalpels and lasers. In preparation and recovery for any medical procedure, no amount of automation can replace the compassionate care of a human nurse.

Orange juice is sold as a commodity on the Intercontinental Exchange along with cotton, sugar, coffee, cocoa, ethanol, and wood pulp. Globally the primary exporter of orange juice is Brazil followed by the United States although it originated in Southeast Asia. The USA grows oranges mostly in Florida, Texas, and California with the common varieties consumed as Valencia, Navel, Blood, and Seville. Primarily known as a strong source of Vitamin C, or ascorbic acid $C_6H_8O_6$, oranges also provide potassium and folic acid; in this one type of fruit you can ward off scurvy, assist your electrolyte balance, and aid in your body's production of red blood cells. They belong to the genus Citrus(one of 160 subcategories), family Rutaceae(typically have a strong smell), order Sapinadales(woody with compound leaves), subclass Rosidae(flowers have many petals), class Magnoliopsida(has two initial baby leaves as a seedling), division Magnoliophyta(it produces flowers), of the kingdom Plantae(it is a plant). Although grossly oversimplified by this sample of information, what is available covers the areas of economics, geography, nutrition, chemistry, botany, and more. The word "orange" is represented by approximately 145,000,000 internet search results in various aspects including the fruit, the color, a city in Southern California, and a dozen businesses.

A painter's palette has new meaning thanks to the diligence of Pantone and the infinite agility of Photoshop(Adobe) and Paint(Corel). It may be true that the toxicity of paint has been replaced by the prevalence of carpal tunnel but those are the breaks. Mixing dabs of color on a thin wooden board is kind of fun but seeing the pages of a Pantone Guide with virtually every tint of every color imaginable is like opening a metropolitan phone book for the first time. It is a countless array of colors at your fingertips to use in a computer program that allows you to manipulate elements of a picture in a Godlike way. What could Leonardo da Vinci do with an insane amount of paint, charcoal, pencils, paper, and canvas?

Photography has already been revolutionized in the span of a few years by the shift to digital cameras. Sales of film have fallen through the floor followed by the companies whose revenue depended on it. A hundred year old industry overnight went from chemicals on paper to microchips embedded in a plastic postage stamp. Besides being reusable, each memory card can hold hundreds if not thousands of images. This makes photo albums obsolete as something grandma has on her shelf from "a long time ago." As a novelty you can order a custom printed book online from a printing service or you can print just the images you want to put in a regular picture frame. Otherwise every picture you own can be viewed on a television, computer monitor, or digital photo frame. A budget computer can handle the storage, cataloging, cropping, and color adjusting of a lifetime of photos.

What are your questions about life, love, woodworking, kombucha tea, Nascar and the 1942 movie Casablanca? By the power of the internet, you get to have the collective feedback of millions of people from all walks of life. You can post questions on a forum about Nascar on a fan site, ask questions about stock cars on a portal site like www.yahoo.com, or go directly to www.nascar.com to check their FAQ(frequently asked questions) sections. To get an answer to a specific question you can simply type it right into the input field of a search engine like www.google.com or www.ask.com. There are also what I would call encyclopedia sites with information sorted by topic like www.howstuffworks.com, eng.wikipedia.org, and of course www.britannica.com. Of course you can also go straight to the web site of your favorite driver to send an email asking him/her to visit your home town. When you really feel like channeling your inner stalker, you can also research someone on their MySpace, Linkedin, Tumblr, Google+, or Facebook page too.

Record players may be nearly extinct but radio still lives

on. Satellite radio subscribers pay to hear their stations delivered to them via a car mounted box, a portable unit with earphones, or through their web site. Every other radio station in the world still lets you listen for free over the air waves and many of them stream it at the same time from their home page. Another popular practice online is how they let you listen to interviews, shows, and concerts from their searchable archives. From a technical standpoint there are a dozen programs available that let you record anything you hear to a file for later. Although music companies spend hundreds of thousands of dollars working on software solutions to prevent people from saving a copy, a cheap audio cable that connects your input port to output port lets you "...tape it off the radio."

During a prolonged blackout, major hurricane, or earthquake, radio will still be the easiest thing to establish and maintain. It is simple enough for anyone to use regardless if you are a child, someone who shuns technology, or someone who fears it. Restoring, finding, or setting up an internet connection in a troubled area demands some notable resources. In a poverty stricken region, an undeveloped desert, or the harsh conditions of a war zone, radio is king for dispensing news and information quickly.

Skin care for the sake of vanity is much more profitable than for health alone. We prioritize what someone or something looks like over the integrity of what is underneath. This flaw in our internal logic is something marketers will readily exploit to promote a product. Television commercials, billboards, and magazine ads are the perfect media to sell the newest eyeshadow, red lipstick, and age defying wrinkle cream. At first it would seem like the internet would be an even better way for companies to show off their products using glossy interactive web pages to draw in the consumer. However, the intrinsic nature of the world wide web enables buyers to talk to each other as well as directly back to the company. One can conveniently find recall notices, comparable items, detailed reviews, a cost breakdown, a formula history, and a dozen sellers for any skin lotion on the market. In an information rich world, the advantage belongs to the customer.

And there is Sex.

The sheer amount of sexually oriented material available

on your computer in terms of both quantity and variety is staggering. A quick search to find something arousing may be so overwhelming that it becomes either repulsive or academic. Stories range from the classic style of Lady Chatterley's Lover and amusing contemporary fan fiction involving Star Trek characters to role playing scenarios exploring the human psyche's dark fascination with kink. Naturally you can find pictures that cover every spacial position, body type, age preference, racial background, and *ahem* accessory imaginable. Sales of adult magazines such as Playboy, Penthouse, and Hustler have experienced significant declines in recent years as a result. Proliferation of high speed internet at affordable prices has made the stable delivery of erotic video content(porn) akin to turning on a faucet.

As much as you want to peruse the realm of <u>beautiful agony</u>, your employer, parents, teachers, spouse, and religious neighbor would rather you not. For their sake several companies sell software that prevents you from visiting those types of web sites like a moralistic traffic cop. While sexually explicit material may be inappropriate at school, work, and near children, it expands the experience for consenting adults. In isolation it fuels a lonely individual's antisocial tendencies and drains the capacity for meaningful thought. In a healthy relationship, the cornucopia of ideas on the internet can inspire intimacy, laughter, and romanticism.

Finding the exact kind of pr0n(slang for you-know-what) you want on the internet is another unique problem. How do you filter through all the advertisements and annoying pop up windows to get to the smut you actually want? One way would be to waste hours and possibly days sifting through random sites until you have a small list that is worth bookmarking for later. Well, in the non-sexual consumer universe there is a monthly magazine called Consumer Reports which reviews products on a continuous basis. On the internet Jane Duvall has created a very thorough online guide to web sites of a sexual nature which would make CR proud. Since 1997 her site has provided a friendly and informative atmosphere to learn more about this universal subject. Jane's Guide is easy to find and it is best characterized by a stick figure self portrait on the home page.

Both telephones(land line) and tape recorders owe their demise to the versatility of the personal computer. Obviously the tape recorder exists nowadays as a garage sale staple since many portable music players(read MP3) and cell phones have a recording feature as well. A growing generation of music lovers have also never even seen a cassette tape on sale at the local record store; no, there are no more of those either. Oh and it is practically a challenge to find a home computer that lacks recording capability. A microphone jack has been standard issue for years even though the computer manufacturers seemed to view microphones themselves as an optional thing to sell you later. Once you get around to buying a mic, you can tell your grandkids how you helped to make telephones obsolete. Using any one of a number of software programs, you can speak with other people through their computer at a cost that makes the phone company cringe. VoIP(or Voice Over I-P{internet protocol} as the techies know it) is the process by which the computer sends your voice back and forth across the internet as data packets. Your vocal vibrations turn into analog electrical impulses that are translated into the computer's native language of ones and zeros.

Let us take a deeper look at this mutant love child of entertainment's altar and the office's former workhorse. Picture a hundred years of television deliciousness joined with the phantom sounds of a billion pages click clacking on a swarm of typewriters. Jack Benny, Rod Serling, Archie Bunker, Tony Soprano, and Lucille Ball coexist on the same machine as 1040 tax forms, late night term papers, meaningless office memos, and the last will and testament of Elvis A. Presley. Instead of checking the couch potato's bible for when your favorite show will be on, you can choose exactly what you want from the mystical TV guide called the internet. The sampling of episodes available right this second demonstrates how the entire catalog of everything, from the multitude of countries, channels, and broadcast dates is inevitably available. In case you have not noticed, Wite-out, carbon paper, and correction tape have left the building, literally and figuratively; companies still make them for a the 19th century holdouts who know it is impossible for mankind to walk on the moon. Conveniently for the writing impaired, the modern word processor has a built in spell check feature, thesaurus, and a few dozen included fonts(which can easily be expanded to a few thousand); a manual typewriter has one typeset only, poor thing. In a testament to those able to multitask well or simply have ADD tendencies, a personal computer allows you to split the screen so you can write down your innermost thoughts and watch the "Oh my nose!" episode of Brady Bunch side by side.

Underwear is a universal source of emotional triggers for all but the most primitive of tribes on Earth. It is well documented, discussed, researched, flaunted, and sold on webcams confessions, random blogs, fetish chat rooms, photo galleries, and online stores. Plain white cotton panties and tighty whities are associated with a degree of innocence, cleanliness, and a state of boredom. Perform a quick keyword search and you will yield a historical time line, popularity by country, and sales figures by age, racial background, and gender. The words lingerie, Mormon, frilly, breathable, discount, and boxers will each add a separate element to further filter the results of your search. sing a highly specific set of words like "blue frilly hemp discount boxers striped used underwear" may reduce your 237,000,000 items to a "no items found" message, a couple of listings with exactly what you wanted, or an excerpt from this book referencing this chapter. Repeat the same search in a few minutes, a few days, or a month from now and someone may have added the information you were looking for.

After winning a brutal war against the smaller and yet superior Betamax, VHS tapes are losing their empire to the cruel dustbin of history. The mass adoption of the DVD may have pushed the VCR(video cassette recorder for the kids out there) to the lower level of the entertainment center but the advent of broadband internet is kicking it out the door. To the chagrin of the movie companies, people have been repurchasing the same movies on DVD they have had as tapes because of the better picture quality and space savings. People still keep a VCR around either to record television shows or just to play back a wedding video or home movie they have collecting dust. As it becomes easier to simply watch your favorite shows off the internet or use a DVR(Digital Video Recorder box courtesy of the cable company), the VCR is losing its key selling point. When a major retailer like Target or Costco decides to offer a drop off dubbing service where you can leave your VHS tape and get a remastered DVD a few days later, it will be time for the VCR's epitaph.

Whether the weather forecasts are absolutely correct is still a debate for the armchair meteorologists but they are close enough for me. Five day forecasts can be sent to your cell phone, PDA(now pronounced "tablet"), and personal computer at your convenience. Severe weather alerts can reach you even if you are at a remote location away from the traffic and lights of the city. Satellite tracking of storm systems has become as widely accepted as looking a calendar to check the date. Accurate predictions of "Acts of God" type events like floods, earthquakes, hurricanes, and volcanic eruptions cannot be made yet but the processing power of our computers are moving us closer to the day.

A trip to the dentist to receive an x-ray used to involve wearing a leaden apron and while he discreetly takes the picture from another room. Nowadays the sensor fits in your mouth while the source shoots you with a fraction of the energy as before. A brief second later the image appears on a full sized computer screen where he can save it along with a few hundred of his other patient profiles. At some point my dentist will be able to tie in my medical records with those from my regular doctor. Like the fabled permanent record that you get threatened with marring as a kid, this detailed history will be directly useful to you. One day you and your healing professional will be able to readily access an interactive diagram of every inoculation, medication, condition, ailment, and treatment you have ever experienced..

Immortalized as part of the industrial age, the yellow pages had become synonymous with finding a particular business in your local community. Anyone who wanted to thrive by more than word of mouth would absolutely place an advertisement in the telephone book. Every year the phone company would drop this huge and handy book on the doorstep of every single house, apartment, and business logistically possible. In the last few years, mine has immediately gone to be recycled minutes after it arrives at my place or become a strongman prop to test my tearing strength.

Thanks to the internet you can not only look up the phone number, hours of operation, and address of a new restaurant, but your computer will also map out the directions, show you reviews, and display a color menu of their tastiest items. For the gypsy minded and those willing to travel outside their neighborhood, the internet further mocks ye olde telephone book by providing the same access to any city in the world; during my childhood, the local library used to have a room full of telephone books for this same purpose.

A zoo has the expressed goal of educating its captors about creatures that would normally prey on them in the wild. From the safe distance of metal bars and man made pits, visitors can observe nature's deadliest animals from afar. With the assistance of the internet you can go one step further. Take virtual tours of zoos around the world by exploring on location videos, three dimensional walkthroughs, and listening to dramatic narratives. Of course the most dangerous animal you can view form a distance is still the human one. Looking at the Earth itself as a massive habitat, you can definitely use the computer to watch how Homo sapiens fight for territory, build their colonies, and play games with each other.

Now that we have covered a plethora of things from A to Z, there are still a few thousand things yet to discuss. From this box of thematic chocolates, you should know there is no limit to what the computer can potentially do. Stretching your rarely used imagination beyond the point of "I didn't know it could..." is what you need to do. Even if the means to do what you want does not exist, kindly believe that it will soon.

Libraries exist for the purpose of sharing knowledge resources with the community and allowing us to freely learn more than we could by our own personal experience. To take that a step further, the computer's role as an infinite library improves our lot in life by enabling us to do infinitely more. It extends our eyesight to the edges of the galaxy and to the insides of an atom's core. Invisible shelves hold Brahm's Lullaby, the soundtrack to every episode of Buffy the Vampire Slayer, the Beatles' original mastered tracks, the howl of a wild timber wolf, and BBC interviews with the fourteenth Dalai Lama. Like a bottomless toolbox, it provides lasers for doctors, welders for factories, salesmen for sculptors, slaves for engineers, and rainbows for painters. Harold's Purple Crayon is truly embodied by the wonders of the computer because it really does help you create anything you want.

Your Neighbor's Web Site
4

Anyone can have a web page made by them, for them, or about them. Your next door neighbors probably have one right now about their dog Jesse and how he leaves massive poops on your front lawn. They have a daily blog about his fecal habits and the time he puked chocolate vomit the day after Halloween. There are vacation pictures at Disneyland with him wearing a Mickey Mouse hat while he sniffs a confused rookie in a Pluto costume. Jesse's video page has links to Scooby Doo clips and shaky home movies splashing in an inflatable kiddie pool with Who Let the Dogs Out as the soundtrack.

The woman down the street with the five cats and crowded shelves of ceramic cow collectibles has a web cam her nephew Gary set up at Christmas so she can watch her kitties at work when she gets bored. Her home page has I LOVE MY MEOW MEOWS with a camera view of the living room and the kitchen where her sweetie pies eat their supper. At the bottom of the screen, a page counter shows 256,763 visitors since Gary posted a link to Tia's cats on his Facebook page last month.

Hundreds of miles away someone has created an online shrine to a cartoon ferret named Edgar, video clips of Nigerian tribal songs combined into a reggaeton mashup, an interactive shopping experience for a fantasy wine cellar, and obscure artsy montages made from 1970's milk cartons. If there are none yet, wait a few weeks and there will be.

* * *

A web page starts from a blank sheet of paper in someone's mind. On this scrap of paper you can use any kind of pen, pencil or crayon to draw on it. You can paste pictures on it without using stinky glue or messy paste to tack it down. Although you can only "see" a section of the paper at any given time, it can be wider than a poster and/or longer than a roll of toilet paper. In a simple sense, a web page is a collage, a scrapbook page, or simply scribblings on a scratchpad brought to life on a computer.

There are a number of special effects on a web page that would require some Hogwarts level magic to bring to life on a plain 8 ½ x 11 sheet. Having words move on a page or playing calypso music in the background are basic tricks. Embedding comic book style animation or showing home movies are not that far a stretch for the untold generations raised on a stream of television. For the medieval readers, just consider a shrunken theater stage compressed onto a scroll. For the tribes hearing this from a spoken retelling, it is the shaman scrawling visions in the sand.

From the power of a touch (or the trigger of a mouse click), we call forth another page using a link. This link is a doorway to another page on the web site, a page on anyone else's web site, or simply just a single picture. Like any real magical gateway, we hope that it is labeled properly so we know where we are being taken, much like a good tour guide. There are deceptive links that get you lost, broken links that yield a blank page, and sometimes bad connections with a sign that says come back later (if you are imagining the story about a young blonde girl and a grinning cat, you have the right frame of mind).

This technological phenom is aptly called the world wide web because you can get snared into the addictive nature of what's available like a fly caught in a spider's trap. The connectedness of all these machines literally does form an information web that you can skate across from one point to any one other point, sending or pulling data to and from anyone. Recipes from an organic cooking club in Berkeley, California are as accessible as tourist postcards from Machu Pichu, Peru or the weather forecast for a week at the Vatican. Although the web is only a subset of the internet, it is the most familiar aspect and application of the global network of computers.

Due to manufacturer politics rather than consumer convenience, the small handful of web browsers by popularity are Internet Explorer(comes with every single copy of Microsoft Windows), Chrome(from the Google people), Firefox(this one is the hip rookie that millions have downloaded for use willingly), Safari(comes with every modern Mac), and Opera(another up and comer with grass roots support); others do exist off the bragging rights radar including ISP branded browsers and the summer project by your local university computer student. Profitability is not a factor since this piece of software has always been free or bundled with your paid internet service à la the old gift-with-purchase practice. All of them serve the same function and possess a laundry lists of features that overlap to a certain degree. However they each have a slightly different appearance, use different amounts of system resources, and have different levels of vulnerabilities to the bad things(viruses, adware, spyware, and simple crashing for no obvious reason). Be sure to also check their home page every so often to make sure you are using the newest version available; the auto update feature does take of that but I sometimes turn it off to nurture my illusion of control. In the grand scheme of existence, your web browser choice is a little more important than the shade of lipstick your aunt Anna wears at her next funeral but less important than having enough fiber in your diet.

By typing in the ip(internet protocol is what it stands for but this will not be on the quiz) address of a particular web server into any web browser you will get the web site of your choice. An IP address is composed of four numbers separated by periods and lucky for the 99.9% of the population that do not have a Rain Man's memory, the internet was set up with us stupid humans in mind. Specific computers on the internet are established as DNS(domain name server) servers with the sole purpose of translating that www.yahoo.com or www.stanford.edu location to the numeric address that the traffic monitors actually recognize. Typing in the home page address and hitting enter will get you the index file of the main directory by default; it is the host of a party that always greets you when you first walk into the house. When you see something like www.hell.com/engine/travel/flowers.jpg the slashes represent subdirectories, or rooms within rooms in the house of www.hell.com; the "engine" room has a room inside of it called "travel" with a picture titled flowers in it(the Russian doll within a doll analogy also works well to explain this if you happen to have one handy).

Essentially a web browser is a file grabber that mainly lets you download text, picture, sound, and video files for viewing

59

and send back requests for more. This is your news, sitcoms, movies, radio, horror flicks, and romance novels combined into a single universal interface. [Cue Journey song] "Any way you want it. That's the way you need it. Any way you want it..."

Thanks to a few restless minds in Silicon Valley a browser can also run small programs using the coffee run screenplay moniker javascript or the fresh faced HTML5. Simple games and watered down versions of popular desktop software are the common application of these protocols. Programmers must craft lite versions to satisfy the impatience of internet users with even the slow connection.

The address field shows both what you are looking at and where you can type in what you want to look at next. It is the computer's path to open a directory or show the file you requested. Throw in **C:** in the field on Firefox or Internet Explorer and it will show you the contents of your primary hard drive; clicking on the folders is another way to search through your hard drive as opposed to going through the little picture of the drive. Should you find yourself in the careful hands of technical support and they claim you need a super secret update not available on their web site, they may give you a path to their server using the ftp command; it will read something like a web site address but start with ftp as the prefix. For all the web surfing time you have logged, you may have already noticed how the humble http:// in the address field is a persistent fixture on the page; if you forget to type that part in, the browser will know what you mean even if decide to skip those seven keystrokes.

Assuming you are connected to the internet, even typing in gobbledygook into the address field will not leave you in a lurch; seriously, Llanfairpwllgwyngyll is a real place. Your default search engine will try to find out what you meant and take you to the closest match. If you know the exact spelling of your destination, be sure to eyeball it carefully when you manually type it in lest you be directed to a shady counterpart who preys on your poor dexterity; there is a cottage industry relying on page view ads to generate income whenever your poor spelling prowess takes you to http://besybuy.com/, http://www.google.cm, http://www.macyss.com, http://www.myspace.com.com, or http://wwwaol.com. Should you get most of the address correct(when you are trying to reach a certain section on the site) you may just end up with an unknown page warning on the web site you wanted. Getting the "Server not found" or "The page cannot be displayed" message means your internet connection has been severed or just does not exist.

When you see the https:// prefix during a sensitive online transaction, it means a security system between you and the web site has been initiated; the "s" on the end of the http(hypertext transfer protocol not Holy Typing Thunder Pants, a loud invocation engineers almost never use) is the giveaway. The standard safety protocol working in tandem is a 128 bit SSL(Secure Socket Layer and if you are smarter than a fifth grader you would have known) or its successor TLS(Transfer Layer Security which you can ask Sheldon to explain to you). Down at the bottom of the web browser you will also see a padlock symbol to confirm your encryption status is active. Any time you submit your credit card data, bank account numbers, or any other type of financial information to a web site, those safeguards should be in full effect.

To satisfy the tragic demons of paranoia, bear in mind nothing except abstinence will protect you a hundred percent of the time. For the average citizen shopping online, this is the flak jacket which protects you from getting your data stolen. Of course you can get phished to a secure phony web site much like you would if you walked up to a fake ATM machine but the convoluted IP address will give it away. Someone who has read up on encryption protocols and wifi piggybacking potentially could sniff out your wireless signal when you are at an internet hotspot and monitor everything on your hard drive; the odds are just like a head shot in a drive by shooting... with the exception of being in known hacker territory, a sniper would have to target you specifically. Instead of the "XX days without an accident" sign you see at work, you can tape up a sheet of paper on the back of your monitor which says " ___ days without being hacked."

For an extra credit assignment you can brainstorm the implications of using a proxy server for privacy, mischief, security, and taking a Tor of the internet. Before you pull out your pocket guide to proxies, think back to those movies where you see a room

61

full of people trying to trace a lone hacker who has routed his signal through computers all across the globe; the older the film, the more likely it is going to be a scruffy looking guy dialing directly into the server from his parents' basement. A proxy is a stand in, a middle man, a substitute for the main guy, which in the world of computers can be linked virtually ad infinitum. Using a set of easily learned recipe of instructions, you too can employ a web proxy to surf anonymously, using its IP address as your gateway to the rest of the big bad internet. Better yet, this is the way industrious young scholars circumvent the censorship imposed by Orwell's Big Brother.

Speaking of servers, you are going to need one of your own if you are going to join the cult of the internet. In order for you to become recognized as a card carrying member of the connected, some server space is essential even if you have to buy, lease, borrow, beg for it, or endure annoying sponsorship ads. Your pretty profile on Myspace, YouTube, eHarmony, Flickr, or Facebook does count as a gleaming billboard on the information superhighway but it is about time you had a URL to call your own.

On the frontier of the wild open cyberspace, you are going to have to pick your domain name before you get to show off your soul. The savvy segment of the younger generation casually sports a web address on business cards, resumes, portfolios, and occasionally tattooed on their precious buttocks. Most common nouns, adjectives, verbs, proper names, catchy phrases, popular acronyms, and other low hanging fruits of speech have already been used or are inhabited by squatters hoping to make an easy buck. Fantasy words, nonsense blurts, terse profanities, clever anagrams, and alphanumeric refugees from the phonebook round out the current crop of website names. Coming up with a virgin URL is going to require a stroke of genius, a dash of blind luck, and a masochistic sense of patience....or a wad of cash to buy someone out.

To host your online party, you are going to need virtual room for the dance floor, open bar, snack area, hookah lounge, and outdoor seating. Spending money for your monthly hosting grants you an empty warehouse of space and a designated amount of data traffic to the server. Scripts for a chat room can be loaded so users

62

can mingle in an endless verbal banter. Discussion forums or blogs may be installed onto your server with no additional cost as part of your hosting package. A shopping cart can easily be linked to an ecommerce vendor on an external web site, much like subcontracting a caterer for the food services. Galleries of videos, pictures, and songs are as easy to share as moving a folder from your hard drive to the server in a single drag and drop. Ambitious coders can hybridize the fancy functions of other web sites using APIs (known to the keyboard set as application program interfaces) to create remix/mashup of maps, demographic data, massive storage buckets, photo libraries, event calendars, wine lists, and custom radio stations; in the real world we would call it partnering with another business to create a mutually beneficial relationship.

Your next step is to get the .com domain and/or the one(s) that coincides with your business plan or astrologer's favorites; the .com(means commercial) was one of the original top level web domains and is top dog on the popularity heap. Settling for .org, .net, .biz, or .info when you really wanted a dot-com is one of those sobering decisions you regret yet ponder every time you hand someone a business card. If you have the clout to lobby ICANN(the almighty yet nonprofit entity Internet Corporation for Assigned Names and Numbers) to create a vanity suffix just for you, it can join a growing club which includes .jobs, .travel, .museum, .mobi, .name, .gov, and .mil; admittedly the last two are USA centric because much of the groundwork was laid by the Americans. To the generations of school kids who memorized the state abbreviations for postal code geography, the youngsters today can spend their time with the list of two letter domain suffixes for the countries of the world.

Of the hundreds of registrars accredited by ICANN to service your needs, odds are that you are probably going to pick the last one you saw advertised. Although the annual fee to lease a domain name originally started out as $50, it can now range from zero(through the old gift with purchase trick) to around fifteen smackeroos. Are you a bargain hunter when it comes to buying toothpaste or are you prey to the pit bull of convenience?

Once you have completed your requisite "paperwork"

there may be a spark of deja vu to the time you applied for your first car loan. Your mother's maiden name and blood type are not required yet but the verification process may evolve to that one day. Through the fabulous whois database (almost like the white pages of domain servers), you can look up background information on any web site on the planet; ICANN can tell you more about it on their internic.net site. At the very least the records will show the site's expiration date, creation date, name servers, the registrar used, and a miscellaneous status line. On a good day, you can also peruse the registrant's real name and address, the administrative contact information, and their technical person's vitals.

Go on. Hang up your shingle and open for business. Tell the world you have your magnificent web site up now. People from Bangladesh to Buenos Aires can type in your URL to bask in your home page. All the time you spent planning on a memorable name, developing a hardy business model, designing a clean layout, choosing a color scheme, and paying those programmers in India, China, Russia, Romania, or the Ukraine are about to pay off(go ahead, admit you planned to clone a web site to use as a template because housing contractors make tract homes all the time).

How will people know to go to your web site? Post flyers with your web address up at the local library, coffee shop, nail salon, gym, and church bulletin. Hire those hardworking mailbox stuffers to slip business cards underneath windshield wipers on cars at the supermarket, shopping mall, and swap meet; leftovers can be left on people's doormats at home or wedged in their doorways. Take out ads in the local newspaper, on the backs of real estate booklets, bus benches, the sides of buses, billboards, restaurant menus, and in every magazine at the checkout aisle. Get your street team ready to staple fortune cookie strips of paper to telephone poles with reckless abandon. Place those television commercials on heavy rotation during the superbowl, prime time, Saturday morning cartoons, season cliffhangers, and right before each of the late night talk shows.

The other theory is what goes on the web stays on the web. Cross pollination to the virtual world is obtrusive when you are enjoying a nature hike, savoring a meal at a fine restaurant, feeding ducks at a local park, or relaxing at the beach. Many diehard moviegoers remember when the product placement trend in movies began, taking its toll on their cinema innocence. Unfortunately, the prevalence of advertising on and about the internet tells us the flow of money is still primarily about eyeballs.

Go ahead, whore your web name to random strangers every which way but lose the idea about bulk email advertising; we *will* curse your name in the dark of night. Otherwise it is your solemn duty to trade banner ads with other complementary web sites who can drive traffic to your web site. If you have to pay for advertisements to your site because you have no online friends make sure the annoyance factor is low. Based on the dwindling attention span of the average internet addict, spiking your ongoing online marketing budget may be a waste of dough.

Of course you can pray for people to passively find you through the help of your favorite search engine. Monkeying around with keyword proliferation within your site may or may not drive your search engine ranking up higher when you are googled(it has already entered the language as the generic term for an internet search despite there being several other engines out there); the SEO(search engine optimization) specialist has joined today's workforce as its own employable position, coming soon to a trade school near you. Luckily the niche your business fulfills should be profitable when just one tenth of a percent of the people who see your web page buys something.

The alternate way to drive people to your little corner of the web is to work on your SMO(Social Media Optimization is the buzzword phrase). Instead of waiting for people to look for you, your highly paid marketing army talks about your product, service, and company on Facebook, Twitter, and other social networks. If you know what you are doing, it is a target marketing wet dream because the niches are easy to engage. Holding tightly to a traditional spend-and-wish advertising model and refusing to learn the new customer landscape is culling companies faster than a bad date with Wall Street.

Once they land on your site, the user absolutely must have the latest and greatest plug-ins, add ons, and updates to

65

experience all your pop-ups, animations, and flashy effects. Bleeding edge on the web means you alienate anyone who is easily confused by the whole "you need BLANK to continue" nag screen. Accommodating everyone involves a bland trade off unless your web coder has included a hidden diagnostic prior to navigating through your site. Your basic browser is a pretty brilliant creation but the bonus features are an exercise in creative chaos.

Streaming video is a perfect example of something people universally crave but often have to troubleshoot a dozen ways to watch it. Whatever browser you use will require a full configuration similar to how each car you drive needs the steering wheel, air ventilation/temperature, seat, mirrors, and radio stations adjusted to your personal preferences. During the setup of any plugin, the invisible wizard will ask you how fast you connect to the internet, as if you are an expert on that sort of thing. From the big shelf of video players you get to choose between Windows Media(using ActiveX), Adobe Flash, Apple Quicktime, Real Media, DivX, Miro, and whatever happens to be popular around the water cooler; most of the players mentioned will happily play most types of video files most of the time if you are lucky. Normally you will install what the computer tells you to install, like a good little boy or girl because it says it needs it. Moments later you discover its has hijacked a slew of other file types since you let it interface with your computer. The longer you spend marking time online, the more likely your pitiful computer has the accumulated attic of software junk. On your springtime desktop clean up, keep a few of your favorites and orphan the others in the spirit of leanness.

For the rare breed of literati who enjoys reading more than a news feed headline or blog posting, she must sort through the undocumented rules of document plug-ins; although electronic scrolling is like reading an unraveling paper towel, this annoyance can be avoided on a tablet by tapping or swiping through the pages. The commercially popular Microsoft Word(.doc) and attorney's choice Corel Word Perfect(.wpd) files will confuse your computer unless you shelled out the cash to have the respective programs to interpret them; however if you brilliantly installed the **free** open source Open Office, it would take care of it too. An Open Document text file(.odt) is conveniently read and written by the selfsame bona fide universally available word processor, spreadsheet, presentation, mathematics, drawing, and database suite. Adobe's Acrobat Reader is a more common than saltwater yet if for some reason you are unable to open a .pdf file, their web site also has a complimentary download. Plain text is showcased by the unpretentious unformatted typewriter scented .txt suffix which practically every healthy electric computer can process. Portable electronic book aficionados are typically locked into proprietary file formats which include but are not limited to .fb2, .fub, .gpf, .imp, .kml, .lit, .lrf, .mobi, and .wol; this means buying from their specific store is easy as apple pie but sharing with friends, loaning it to a neighbor, or borrowing from your wife is a major p-a-i-n in the butt.

The plug-ins of Christmas future are probably going to be tools which help to bring the three dimensional world closer to you. A few scattered web sites are showing evidence of walkable landscapes, teleportation maps, full exterior object examination, cross sectional medical exploration, and the customized morphing of your body. Early versions on the table are rough to navigate, cartoonish in appearance, slow to load up, confusing to the average user, and sparse in detail. These first attempts foreshadow the adaptability of the infant web browser in its revolution to replace standalone software installations; as the specific plug-ins are being developed in an software evolution free-for-all, you can listen for the survivors in your favorite weekly tech column.

With the glorious web browser becoming the leading software application in terms of sheer popularity, it behooves us to *bookmark* our *favorites*. A shy vanilla surfer may have barely a handful of repeat url encounters versus the aggressive internet-active soul who has a thousand or so web handouts. Like the bad little addicts we are, we want our fix quickly, savoring the instant each page loads. In our shopping frenzy, we will light up twenty favorite store sites, review pages, coupon blogs, deal forums, and auction links fanned out in tabs. Our news mood cascades over a dozen online television stations, newspapers, financial analysts, streaming field reporters, and independent journalist blogs in a miniature glowing wall of "what's happening in the world now." Oh, the not-so-subtle row of quick links below the address bar also gives away my guilty pleasures of celebrity gossip, food porn, rude t-shirts, lottery jackpot, and music videos on the Tube.

To take advantage of the clusters of other lollygaggers out there wasting their productive hours, it is fun to see what they think is fresh, amusing, interesting, brilliant, and most of all utterly fantastic. It is always a beautiful thing to see the most popular urls on the internet grouped by category. Who knows what you might accidentally stumble upon by checking out someone else's list. Maybe you reddit yourself on the train while browsing for news fresh off the vine and you want to save it for later. If you come across a scrumptious ice cream that is unbelievably delicious, you would want to tell your best friend. Some people will simply want to know what the fark is up with this ding dang here old internet. You digg?

The ultra efficient armchair computer commanders would rather have everything brought to them on a silver platter. Besides the inflatable companionship, pizza, grocery flyers, high pressure religion, and bottled dairy delivered to your front door, you can get up to the second updates from your favorite web sites. Your home base is a newsreader page called an aggregator; no, it is not an alligator, an aggravater, or an elevator. Myriad web sites are already set up to serve your RSS(Really Simple Syndication) or Atom news feed to your personal page. It is like having a bed of nails intravenously inject you with a continuous salty spray of data.

Hyperactive attention deficit disorder sugar bombed caffeine laden normal people prefer to actually engage the community. Every real world social vehicle has an online equivalent, a digital counterpart that smells vaguely familiar...

Blogs are the online diaries where everyone gets a peek; it can even be set to a private mode for your solitary contemplation. Your journal entries may be as dull as leftover day old toast or as tawdry as a call girl's psychological romps.

Forums are bulletin boards where everyone gets a say; its categories are neatly sorted by topic into individual discussion threads. Your pithy posts are time stamped, moderated, mocked, applauded, and analyzed.

Web rings are clubs of independent web sites united by a topic; an earnest administrator creates one site, invites others to join the group, and maintains a running tally of the list. Your arrival at one site on the ring allows you to view the list and go to any other.

Fan sites are literally shrines documenting an obsession; the more famous you are, the more images, videos, sound bites, quotes, and merchandise is out there. Your road to becoming a crazed stalker begins right here.

An avatar is your computer generated doll, totem, symbol, figure, and self; in a given software environment, you can customize its features to look exactly like you or your fantasy face and body. Your trip down the rabbit hole into the world of create belief begins when you fashion your virtual you...

No, you do not have to wander the matrix alone!

Social networks are what the textbook writers classify as the massive pseudo anonymous guiltless computer cults born on the internet. You can invite random people to become your friend without having to go to a special campground, fast for a few days, shave your head, adopt a new god, lose your precious iPad, consume questionable food, or wear weird clothes. In fact, the implicit goal in becoming a member is to recruit as many people as possible to become your online friend. Who needs a close circle of meaningful relationships when you can have a few thousand causal one liners from your online acquaintances? The companies who run these web sites need the advertising dollars just as much as you need the bragging rights of being well connected.

Run with extra sharp scissors quickly through the corporate hallways of MySpace! Have lunch with friends from Friendster, paint some faces on Facebook, bebop down to Bebo, call someone from Classmates, say hello to someone on hi5, hang out at Habbo, look up your boss on LinkedIn, join the BlackPlanet, flip through Flixter, mix it up on Monday with Mixi, and browse a bit of QQ. Tell everyone to watch scenes from Xanadu while they are logged on to Xanga.

When you have overdosed on interacting online with your posse, the best prescription is some good old fashioned television. Keep your wicker basket of remotes right where they are because the window to the world is really at your computer. Satellite and cable TV operators who rely on pay-per-view and on-demand programming for their cash cow are competing against the ultimate whatever-video-you-want-now internet paradigm. The wishing video machine may not yet have everything we want to watch but the sheer availability of every type of video every produced is significant; your infinite library recently opened and the stacks are still filling up.

Online video is only a recent milestone in the annals of computer history. While the bandwidth rich population can readily participate in this boob tube buffet, it is still a technical challenge to feed everyone. There is a huge difference between watching a five minute scene on a quarter of the screen versus filling the whole monitor with a high definition streaming movie. Until the fiber optic future reaches everyone, we will have to content ourselves with palm sized shows.

Who needs the Nielsens when we have Alexa rankings for the web? Typically a TV show will live and die by the number of viewers because the cost to produce a show has to be offset by the ad revenue it generates. A web site is so cheap and relatively easy to produce that anyone can build a popular one to make money.

It is all about popular![Cue Kristin Chenoweth]

Email Rules of Etiquette
5

In the olden days, the sending of a letter was blessedly simple. Pour your thoughts out on a sheet of your favorite stationary or random scrap of paper, fold it up, and put it in an envelope. After you sealed it, you turned it over to put your return address on the top left corner and scribbled down who you wanted to send it to on the center of the envelope. Before you dropped it off at the post office you made sure you licked a postage stamp and slapped it on the top right hand corner plain as day.

A big development in the world of letter writing was the introduction of self adhesive stamps. It was a radical departure from those ancient times when you had to carefully tear out the postage stamp from a page and slide it across your tongue. If only someone would come up with a way to cash in all those miscellaneous bits of random postage that accumulates every time there is a rate change. Or maybe someone could even create a way to send a letter without having to constantly buy stamps, envelopes, or paper.

Poof!

Welcome to the world of email. It was born as a child of the 1960's, hit a massive growth spurt in the 1980's, and has reached every corner of the globe since then. Anyone with access to a personal computer has one or more email addresses from which to send and receive messages. Even people who cannot afford telephone service are able to use an email account through a public library or for a moderate fee at an internet cafe. To this day many companies continue to offer free email accounts as long as you find some way to get on the internet in the first place. Once you have a connection to the big mysterious network, you can send as many letters, I mean emails, as you like.

Imagine sitting at a desk with a bottomless cardboard box to your left. As soon as you write a message and put an address on it, you casually drop it in the box. Moments later it appears on the desk of the person you wished it to, waiting to be read. You did not have to pay for postage stamps for any of the letters you dropped in the box. Of course, as long as your cardboard box was sparkling with pixie dust you were good to go. For those of us who have been using this email thing for years, we do take this practice for granted.

Variations on the theme include instant messaging and SMS(short message service, the name only the brochure uses) or text messaging. The latter commonly refers to the sending of short text based messages between cellphones for when you have to tell someone something but you cannot be bothered with actually calling them. As a monetary source, the cell carriers get you coming and going by charging you to read ones you receive and to send these short missives out. Sure there are bulk rate and unlimited plans for texting available but it has caused our language skills to decompose rather than thrive. A linguistic shorthand that flourishes on SMS originally started by annoying people who frequent instant messenger services on the internet. WTF RT? L8R LOL. The short version of the story is that instant messenging is analogous to a phone conversation while email is more like postal mail.

Unlike snail mail, as the USPS paper stuff is affectionately called, email does not have that pesky pay more to send a bigger package rule. Although you can bounce back an email if it is over your mailbox limits, the amount of words it would require to do that would exceed your physical ability to type in one sitting. Attaching files to your email is just like sending a package with your letter or conversely a mailing label to your parcel. Depending on the size of the pictures you attach, it could be like handing your postman a book or a bowling ball. Trying to attach a full length movie to an email for sending would be like mailing an elephant. My mail carrier would laugh all the way down the street for me even asking to bring that back to the main post office. Your email account will simply give you a boring undeliverable note as an automated reply for trying. To transport the proverbial six ton pachyderm, you need either peer to peer help or some fiber optic cable and a ftp client(file transfer protocol software that lets you maintain a direct connection with another computer for moving files).

"When I was a kid, we had to dial up on a 14.4 kbps modem to connect to the internet and all our email was in a plain text format. Everything was in ASCII and we were happy to have it! We didn't have none of that HTML embedded anything with all those pictures and links. The young people today don't know how lucky they have it."[In my best crotchety old man voice]

ASCII(pronounced ask-ee) stands for American Standard Code for Information Interchange and you will never be asked about the acronym ever. It is the name of a basic text system that contains the lower and upper case alphabet, integers from 0 to 9, punctuation marks, and the rest of the basic symbols we borrowed from the typewriter. There is no **bold**, no <u>underlining</u>, no *italics*, and no **crazy fonts** to distract you from your message making mission. You will also be living with the exact same size lettering no matter what you do. When you see those ubiquitous ReadMe files in .txt format you know those use our lovably familiar ASCII. Some people's immediate reaction is to discount this as old, boring, and as hard to read as the books on the back shelves of the library. Plain text for email is important because it is intrinsically safe from viruses, adware, and other hazards of the computer world. It can only contain ordinary words so the worst thing that can happen is that you may be offended by something vulgar.

HTML formatted email is much cooler looking and prettier by far. Fluffy bunnies, cuddly teddy bears, a field of daisies, a wall of jellybeans, or a picture of a Maui sunrise can serve as the background of your email. For the grim and gritty, gothic steeples, glistening machine guns, bloody fistfights, dark alleys, and prison walls can be placed underneath your words as well. When you see an advertisement link that you can click on to a web page, you know you are using the fun filled HTML style email. Colorful images, glaring bold type, and flashing signs are indicative too. Certain types of file attachments are now shown on the lower portion of the email instead of having to click on them individually to view.

From now on, think of a file attachment as the mysterious package in the brown paper wrapper. Looking at the suffix of the file you can tell what type of file it might be and whether it is safe or potentially dangerous. Your common sense will protect you far better than any automatic email scanning program.

When you see a file that ends in .txt you know it is always safe because it is full of harmless words. Image files such are .jpg, .gif, or .tiff are usually safe most of the time since those types of viruses are rare. Audio files(.mp3, .wav, .wma, .flac) and video files(.avi, .mpg, .rm, .mov) have a medium risk and it is best to have your antivirus program scan anything someone sends you. Program files(.exe for executable) or any type of file you do not readily recognize should be treated like a terrorist threat. This means avoid all contact and delete it with extreme prejudice. If you absolutely must see what it is, save it to a folder where you can have a SWAT Team check to make sure it is safe. Check your antivirus software for updates on the internet immediately and when it done, right click on the file to choose a manual scan. Be aware that even if the scan comes up safe, an undiscovered virus might still slip by to attack your hard drive.

That cliché *death by a thousand cuts* aptly describes the annoyance of junk email. We call it spam when we see a lot of it in our inbox or we say we got spammed when we tell someone else about it. Bulk mail advertisers see it as a blessing since they can reach thousands if not millions of people at a time without paying for postage stamps. Cheap Viagra. Property in Costa Rica. Publisher's Clearing House. Ink cartridges for sale. More Viagra. Porn. Free laptop. Refinance my house. Diet patch. Local Singles want to meet me! Free ringtones. Discount insurance. Cheap Rolexes.

In my email inbox there is an urgent plea from someone in Nigeria asking for my help to free up $70 million and in exchange they will share a portion of the money with me. All I need to do is to wire them two thousand dollars so the bank can process their claim. Oh my God, if I help him, he will share a few million dollars with me. Uh... no. Seriously, this is a dangerous game of fraud that continues to this day. Its notoriety has reached such proportion that it is referred to as a "419 Scam" from a specific article number in the Nigerian Criminal Code. Variations have also emerged based on fraudulent lottery prizes, fake charities, or someone asking for money to help them out of a "terrible" situation.

Another email, which looks like it is from Bank of America, is telling me that someone has broken into my savings account. It says I need to click on the link and go to their web site to verify my account information. The web page looks real down to the 1-800 customer service number, company motto, and color scheme. I smell a bathtub full of mackerel. First of all, I do not have any accounts at Bank of America. Secondly, the web address looks like it has a whole bunch of extra numbers. Lastly, all the images point to BofA's home page on a mouse hover check but the submit button seems to redirect you to a mishmash address. Someone is fishing for a sucker but I have to give him credit for such a good job in trying to Phish me. Instead of getting annoyed, turn up your copy of You Enjoy Myself and have of scoop of some Ben and Jerry's.

Admittedly the free email providers are getting pretty good with their built in junk mail filters although a few of those pesky things are getting through to my mailbox. The cynic in me says that writing them to remove my name from their mailing list is a futile favor to ask since there are no repercussions other than karma. My other strategy is to funnel everything through an email client like Outlook or Opera, Thunderbird or Eudora, and manually create my own filters to screen out the garbage. Occasionally something valuable does get thrown in the bulk mail folder and it is a tedious ritual to skim through the trash before it gets put out to the curb.

The chain letter has gained a new life via the wonders of email. From a technical point of view, it is fairly harmless except for the time it eats up. In a way it does act like a weak strain of virus except instead of a hidden code forcing you to spread it to other computers, it merely uses the pretext of psychology to make you forward it to all your friends. It is a heartfelt prayer circling the globe, a hilarious blonde joke, a warning about toxic plastic wrap, a petition that saves the rain forest, or a scary ghost picture that leaps up at you after 30 seconds of staring intently. Personally the printable BOGO (buy one get one FREE) coupons from my friends are great to receive but anything that reads like an urban legend has me looking it up on snopes.com to confirm my suspicion of another hoax.

Through the years I have probably had a dozen email accounts, forcing me to remember at least five times that many passwords. Each new job, school, and internet service provider gave me a new account to keep track of. All of the web sites that I managed had several email addresses associated with them as well. At the moment I have a personal email account that I use to stay in touch with my friends and family. Another one is set aside for anything related to work, finances, and online purchases. A third one is fully dedicated to contests, surveys, free offers, and anything else that could potentially lead me to get put on some marketer's mailing lists. On a regular basis, I do change the passwords to keep up a certain degree of security.

You use a difficult password for the same reason you have a strong lock on your front door. Using the word "password" as your password is the classic no-no which has been repeated to you since you could walk. Other generic advice is not use a loved one's name, your birthday, your pet's name, a phone number, or something that can be guessed by looking at the knickknacks on your desk. Anyone who uses the same password for all of their accounts is making life easier with just one to remember. Should someone find out what it is, a thief would have a skeleton key to everything you own. Writing all of them down on a small note card means someone could find it. Using a piece of software to manage all your user names and passwords is fine unless it gets corrupted. It is no easy task to have your life both simple and secure.

When you forget or lose your password you can contact the system administrator to reset your information. It is like calling your mom as a kid at school when you forgot your lunch. This is also the easiest way for a hacker to get access to your personal account. Sophisticated software can rapidly perform a dictionary attack by applying the most common words in the English language. The brute force method of trying every word and letter combination in a six digit sequence may take forever even with a fast processor. However with a little human engineering by way of acting and improvisation, a person could get your login info by pretending to be you.

We will dig deeper into your paranoid fears after we chat about accessing email from the mysterious forces of the universe. Your trusty mail server is the computer responsible for sending your email out to the great big and scary internet and safely holding what people send you. It is your personal virtual post office, postmaster, and PO Box all rolled up into one.

On the client side, you can run a standalone program or a web browser to check your mail. Using a dedicated program like Outlook, Eudora, Thunderbird, or Entourage allows you to read messages, create brand new ones, and organize your mess o' mails without the internet. Once you are connected, the software will automatically get all the email waiting for you and send out the stack queued in your outbox. In this case all of your email resides on your hard drive along with your precious address book. Should something catastrophic happen to your computer you may be S.O.L. as the old folks say. With a web interface on your favorite browser, you can access your emails only when you are connected to the internet. This scenario applies whether you use a free email service like Yahoo, Gmail, Windows Live, Aim, or one of the few hundred other providers out there. Those free services also enable you to check email from the domain accounts you have via your web site. Portability is the word that often associated with webmail because you can email from any computer connected to the internet whether it is at home, your friend's house, an internet cafe, or your office. Unless you save an offline copy of a message or two, your email pretty much stays safely on distant servers managed by professionals 24/7.

A perpetual problem for all users, with email addicts noticing sooner, is the menace of too much email. Your email software will slow to a crawl and eventually crash when you get to the point of no return. Searching for that one message you need makes you constantly reliant on the Find feature because you have not bothered to organize your email into neatly labeled folders. Doing maintenance on your email client is low on the priority list until the proverbial engine light goes on. Webmail is not immune either but at least the interface displays a percentage used graph or a "space remaining in your mailbox" alert when you log in. It is nice to get some warning when you are riding close to oblivion especially if you are not really paying attention. For the casual emailer, the size of today's modern mailbox is big enough that you need not worry about running out of space. Of course if you are still saving those funny videos your coworkers sent as attachments, you know the culprit who is filling up your box. Those annoying junk emails which we ranted about earlier tend to be small but are plentiful enough to push you over the limit if your filters are not properly set.

Whatever you choose to access your email account should have a plethora of settings with which to customize to your heart's content. The bells and whistles, pretty colors, and perky sounds are all parameters at your control. It can be both a chore and a challenge to familiarize yourself with the features on a typical email client but there is no getting around it. Web based email has fewer extras but the basic options are going to still be there. Your required learning list should include creating/editing accounts, manually making junk filters, and managing your mailbox. Although there are many great online tutorials available, my foremost suggestion would be to seek out a class at your community library.

Email in the workplace has a slightly different set of rules that your boss may not have explained to you on day one. Every piece of mail that you come in contact with is owned by the company and can be read, tracked, and subpoenaed if necessary, at any time. That email retention policy where your messages get deleted after a certain date not only saves space on the server but it also prevents emails from being dug up which may incriminate the company. Any company large enough to have an IT department would be shrewd to train everyone on the best ways to manage his/her mailbox as a easy productivity boost. Blatant office flirting, active moonlighting, and spreading company information by email are punishable by termination but grossly low company morale does tend to nurture that though.

As a good worker, upstanding citizen, boy scout, or nice girl, there are some general guidelines as to proper email netiquette. These unspoken rules apply to both your personal and professional correspondence since you are just as civil at home as you are at work....right? When you send an email, hopefully you do take the time to think about who will read it instead of CCing everyone who should have a copy. Knowing when to use "Reply" instead of "Reply All" will also reduce the number of people who quietly think of you as an idiot. Sarcasm rarely travels well to a crowd on email because face value is the norm without a telltale facial expression. A message written in anger carries about the same amount of spite as saying it to someone in person. WRITING IN ALL CAPITALS IS HARD ON THE EYES AND IS CONSIDERED TO BE THE EQUIVALENT OF SHOUTING. Using piss poor vulgar language without fucking caring about the goddamn recipient is flat out rude. A persun who doesnt bother to use the built in spelcheck is frankly anoying as well. Fortunately the grammar police are not nearly as strict here in cyberspace. Excessive emoticon usage is not only distracting but in some states it is surely a crime too.

We forget how much easier it is to send an email compared to a tangible piece of mail. The creation and delivery time has dropped dramatically, especially for sending multiple copies out. Pulling down a menu to add a group of recipients is far different from crating a few boxes of individually stamped, filled, and sealed envelopes to the post office. One guy can send the same amount of messages in an hour by email as you could truck to the post office working non-stop for a month. If he uses bulk email software for a month, you would be trading a lifetime in just stuffing envelopes alone.

Between the personal emails, work related emails, newsletters you want, and the junk mail you hate, you have a grand time sucker on your hands. Imagine getting sixty to a hundred emails daily, all coming continuously throughout the day(you have my personal condolences if you deal with more than that). Your eyes register every word in the stack of subject lines and they stay with you subconsciously until you get a chance to take a big mental break. Unfortunately the huge biological downside is that you, me, and the rest of the people in the world still read at the same rate. Reading off of a computer screen makes your eyes more tired faster than say reading a book for fun. That desire for email etiquette may be overstated for the rare person who gets a few emails a week. A typical person deals with a few reams of metaphoric mail Monday through Sunday whenever he/she logs in. At the push of a button, the magic inbox on your desk produces a generous stack of special No. 10 envelopes, literally forever.

My personal goal is to go on an email diet and I encourage you to do the same. Join me in sending out only what really matters to the people who are really interested in reading it. We will remember that a hand written letter delivered by a mail carrier is still infinitely more romantic than an email. Let us watch our diction, our prose, and our consonants for glaring errors that we would never say out loud. When we get unwanted spam, instead of just passive aggressively tossing it in the trash bin, we will report it to the Internet Service Provider who helped send it.

Chatting, Sharing, and Downloading
Mushrooms in the Forest
6

I love mushrooms. They are quite yummy and nutritious. The number of varieties are plentiful enough to spend an ungodly amount of time cataloging. Some are poisonous, a few are medicinal, and some others are right tasty in a nice cup of soup. Although they are typically expensive to buy at the market, picking some fungus in the forest is a quiet little adventure. Depending on your experience and knowledge, you could get high, get lost, find a cure, or even die from the trip.

Downloading files from the internet is much the same way. The things you find may be thoughtfully amusing, surprisingly enlightening, highly entertaining, and of course potentially dangerous as well. Similarly you can buy many of the same pictures, software, movies, music, and books from your local brick and mortar store. However a download version tends to be much much cheaper even before you factor in the savings in gasoline and commute time.

Downloading is a time honored tradition that predates the dawn of the browser wars between Netscape and Internet Explorer; the empire did strike back to overrun the rebel forces but later Firefox emerged as a return to the source. When the first college students gazed upon Mosaic, they saw how a browser could instantly view pictures and text files from another computer in a room far far away. A rebellion of users emerged from the isolated computer labs and expanded to the homes of restless teenagers, presidents, and rock stars alike. In a short span of years, you could download a galaxy of information, literally about every living thing and penetrating every walk of life.

We like downloading things because it is tremendously easy once you get the hang of it. It also satisfies some basic laziness we cultivated from an early age. As babies, we learned how a simple scream will make mom and dad change our diaper. The same system applied to when we got hungry, irritated, or tired as toddlers. Just as we start to learn a semblance of a work ethic, we are hooked by the magic of the TV remote control. Pushing the little rubber buttons made Elmo and Big Bird and Barney and even Scooby Doo appear! (I speculate that our innate desire for instant gratification can probably be traced back to our previous life incarnation as lab rats where we pushed a button to get our food pellets.)

A picture is a picture is a picture. So in the digital world, the original file is no different than the copy. The degradation you see when you make a photocopy of a photocopy does not occur when you make a straight copy of a file. That is the beauty of sharing files with friends and strangers. Your copy is exactly as good as the original.

By making a duplicate of something easily and without costing any money, you get to be like Jesus and his five loaves of bread. Instead of feeding the masses some carbohydrates, you are providing others with knowledge and entertainment. Many would say this communal attitude is good because sharing is the cornerstone of compassion, charity, and democracy. After all the first thing we are fundamentally taught as children, is how to share. Sharing is caring.

Copyright, trademark, or any other intellectual property laws frown on the whole free-for-all paradigm. Traditional business models rely on possessing things of value to sell to others for a profit. The bustling movie, music, book, art, and software factories are structured to make money by selling ideas packaged in a physical form. Industries generate income for others besides the primary actor/singer/writer/artist/coder by employing studio executives, accountants, marketing specialists, recording engineers, as well as the unsung truck drivers delivering treasures to the store.

Counterfeiting, or selling copies without permission of the owner, is an enormous violation of federal and international law. It is a way of cheating creators by skipping all the research, development, and advertising work that went into producing Lord of the Rings or a Taylor Swift album. This half hearted entrepreneurial attempt to duplicate products for sale out of your garage or living room is almost admirable if not for the karmic backlash. Anyone with half a brain and a watchful eye can set up a street corner operation of hustling bootleg movies and music out of a duffel bag.

Obviously it is a risky business for the seller but the buyer is also taking a chance with getting caught or wasting his money on shoddy product.

On a very basic level, a painter has to make money from what he creates in order to feed himself or else the starving artist dies of hunger. Vincent Van Gogh is famous for his both his world famous works of art and his tragic life of poverty. No less notable is author J.K. Rowling whose wondrous stories transformed her own lifestyle from government welfare to great wealth. Somewhere in between the fates of fortune and failure, there are musicians who would just like a living wage.

Along comes the monkey wrench. The fly in the ointment. A world shaker, heart breaker, and money maker. Yes, it is the networked personal computer.

Borrowing a book from the library, loaning a DVD to a friend, or recording a television program fall under what is called the "fair use" doctrine. What it means is that it is generally okay to share stuff for your personal use. Often people interpret it with the reasoning that as long as they are not selling it for profit it is all right. Although my background does not cover the legal nuances of international law, it is probably safe to say no one remotely envisioned the ramifications of the internet. With the connectivity of a modern day computer, the entire world is your close personal friend.

File sharing involves being connected to a network of computers, many of them faster and more powerful than yours. Behind each computer is a person who wants to listen to music as much as you do. He dreams of being able to hear every single jazz piece ever recorded by Thelonius Monk while slaving away at domestic farm reports. A woman enchanted with Joni Mitchell discovers her soulfulness decades after the coffeehouse years. Two little girls hear an old Shirley Temple song at school and want to sing along to "I Want a Hippopotamus for Christmas" at home.

Did you know grade school children share files with as much ease as a game of hopscotch? In fact, they can download a video of other kids in another school jumping rope in less time that it would take for them to get a rope and go outside themselves. This comfortableness with technology does not make them prodigies or even especially privileged, at least not compared to the way the world is shifting. In urban areas of the planet, the adage "even a child could do it" is a near statement of fact.

For the person who shares files or breathes air, the number of ways to exchange content is practically limitless. As restless caffeine driven sponges with a fractional attention span, we constantly demand better ways to communicate. Our desire for instant gratification pushes us to create not just faster ways to volley ideas but cleverer ones as well. We are not satisfied by merely watching a movie at the theater or just listening to a song from an album. Mikey likes to watch videos in the car, at his desk at work, on his laptop at home, with his phone, during a flight using a portable video player, and at any time he feels like it.

The RIAA(Recording Industry Association of America) is a fee based trade organization representing over fifteen hundred record companies. They are responsible for helping to enact technical standards for audio recordings, certifying album sales(gold for 100,000 units and platinum for 1,000,000 units), and lobbying to Congress for laws related to copyright. Their biggest claim to fame is undoubtedly their public battle with downloaders in which they continue to sue several thousands of people for freely sharing music. A person suspected by the RIAA of downloading songs is subpoenaed by their lawyers and usually offered the chance to settle out of court for a few thousand dollars. Consequently, the RIAA has earned a bit of infamy from those who disagree with the legal grounds for their lawsuits, the sheer strategy of suing your customer base, and the general argument that decreased sales are due to file sharing.

The MPAA(Motion Picture Association of America) is the group representing the movie studios. Their most visible actions are in the implementation of movie rating(G, PG, PG-13, NC-17, R) and the ubiquitous FBI warning displayed on every video tape and disc you can remember. Their efforts to fight movie piracy have been a little less litigious than the folks at the RIAA. Instead of suing the pants off everyone, they are focusing much of their efforts on legislation to target universities, software companies, and web sites which facilitate copyright infringement.

Should these two group become too overzealous, the nonprofit EFF(Electronic Frontier Foundation) located at http://www.eff.org is working to protect people. Their mission is centered around preserving your individual digital rights in this new world of computers. As a proponent of electronic privacy, technical innovation, and free speech they step in to help in many legal cases. Browsing through their archives is a quick way to get a feel for your personal freedoms in the virtual universe today.

While most of those battles rage above our heads, we just want the internet to go faster. Dial up was fine for those early days when all there was to see was plain pages with a picture now and then. Flash enabled pages with animated cartoons, elaborate photo galleries, and streaming music are okay with DSL(it stands for Digital Subscriber Line, another forgettable acronym); it comes from the phone company or a vendor who leases out their phone line. Cable internet is generally faster and comes through the same coaxial line as your TV shows; watching video is normal and web pages blink onto your monitor. Should you be so lucky to have FIOS(FIber Optic Service) available in your area, that is the closest you will be to plugging a cord into your brain. Downloading a song on FIOS or cable takes a few seconds compared to a few minutes on DSL. A movie procured from the internet will be done overnight with DSL. With cable, it will be completed in less time than it takes to watch it. With FIOS as the future, let us just say that trading movies will become like when people traded baseball cards.

Alternative ways to connect to the internet include satellite uplink, leasing T1(or greater) lines, a few developing cellular services(3G, WiMAX, LTE), and piggybacking off the free wifi at a public hotspot. That covers those of us who live in a remote area, have a serious business need, and anyone too cheap to pay for their own internet. Your homework is to research the cell phone options on your own because they are evolving too quickly to track in a printed book. With the exception of the hard line, your speed and reliability depend a lot on your proximity, geography, and equipment.

Suppose you have your internet connection taken care of,

then all you need is some decent peer to peer software to begin. P2P is the pop geek name for a program that lets you share files in real time with others. Sitting around a campfire telling ghost stories to each other is a peer to peer activity without a computer. Now imagine it is a few thousand people in a football stadium tossing CDs from the bleachers to the nose bleed seats; you suspend belief in that your throwing arm is ten times better than the hot dog guy and you never lose your original no matter how many you give away. Multiply that by a few million people with everyone bringing their own personal collection for the massive potluck. That is the power of a peer to peer system.

Napster was the first peer to peer software to scare the bejesus out of the music industry and delight music fans worldwide; Shawn Fanning earned his place in internet history with this creation. With the speed of greased lightning, a user would type in the name of a song or artist and several dozen or more results showed up. In the next couple of years, it is probably safe to say an unknown number of (millions and millions) songs found new homes on people's computers. On a hot summer day in July 2001, the first incarnation of Napster was quietly shut down to comply with a legal injunction.

In Greek mythology, there is a serpent like creature which Heracles(the Roman and Disney version call him Hercules) fights as part of his Twelve Labors. The Hydra's formidable nature is demonstrated by its many dragon heads and how cutting one off would make two more grow in its place.

Following the demise of Napster, the world would shortly see a number of other P2P networks crop up as successors. Bearshare, Morpheus, Kazaa, Limewire, Grokster, iMesh, eMule, Kazaa Lite, and Napigator are just the first nine programs that come to mind as part of the next generation(a given peer to peer network could have more than one software interface to connect to it, like riding different rails cars on the same track). These P2Ps would collectively surpass Napster in the number of users, the number of files shared, and the types of files being shared. Now you could find movies, TV shows, books, software, and pictures in addition to a plethora of music.

Up until this point, your P2P download speed was determined by the number of people who had the file you wanted and your respective connections. It could be pretty zippy when a lot of people had the file you wanted and you could watch the download time super shrink from hours to minutes to seconds. On occasion it would bog down if say a few hundred people wanted a particular file and only a handful had it.

Enter Bit Torrent. With this Peer to Peer software, people who are in the process of downloading a file would immediately start sharing as well. Anyone who has a whole copy of a file is a seeder and someone with a partial copy is a leecher. Dynamic sharing by both leechers and seeders alike increases the overall throughput. *The more people who share a file, the faster you get it.*

Thank you very much Bram Cohen.

Bit Torrent's efficiency represents an absolute miracle for music lovers, movie addicts, and crazed TV fanatics. A musician's complete discography can take less time to obtain than to watch an episode of 60 Minutes. Crisp quality movie files are available weeks before the DVD comes out in the store and occasionally days after it is released in the theater. High definition TV shows are accessible anywhere the internet flows freely regardless of your time zone or the planned broadcast date in your country. As long as it was already shown on the air somewhere, the internet becomes a kind of universal VCR, Tivo, DVR, or PVR.

Before you go off joining a swarm to download the entire Smithsonian archives, Billboard Hot 100 for the last 50 years, or the latest Oscar contenders, be aware there is a showdown coming. Some form of Prohibition is coming down the pike to try to stifle the tendencies of human nature. A number of corporations want a digital martial law to control their streams of profit in the only way they know how to do business. Certain governments want to restrict your access to information to prevent you from knowing the harm a small handful of them are doing. In the name of religion, the restriction of outside ideas is often considered reasonable for believers and those they wish to convert to their way of life. Censorship is such a convoluted proposition for a global village that can instantly talk to one another.

An alternate way for anarchists to communicate on the internet is through the age old Usenet. While the internet is the universe of connected computers, the galaxy of servers which trade posts on these virtual bulletin boards are called the Usenet. With the help of a news client software a person can both read and post items up on various discussion areas. These newsgroups are best characterized by the down and dirty way you can quickly pin your message to the vast cork board in the ether.

Then there are your chat rooms. Clandestine conference rooms where you can talk in private with a special someone or a dozen strangers. You get to volley typed messages back and forth in visual anonymity like an elite TDD(Telecommunication Device for the Deaf) user. Shorthand emoticons, or emotional icons, first festered in their tiny little windows before creeping their way into our written language. Although AOL probably wins the contest in media mentions, there are several other instant messengers including IRC, Yahoo, QQ, MSN, and ICQ time sucking millions around the world.

When you are not dodging pedophiles, angry loners, automated porn texts, idiots, and madmen, you might find real friends online. Besides the emotional rapport of connecting with someone who shares the same interests, you also get to share your favorite pair of shoes. Maybe not your red Jimmy Choos per se, but those pictures of hidden Mickeys you took on your last trip to Disneyworld have got to be shown off. Old Elvis tracks pulled off a cassette tape may not be appealing but you listen daily to the lovingly remastered Janis Joplin concert you got from the sculptor in Phoenix whom you have never actually met. What could be better than faceless virtual support groups to provide you with validation when your wife, husband, girlfriend, boyfriend, parents, boss, and coworkers just do not understand you? At any given time, it is just like mingling blindfolded at a twenty four hour club full of random rooms, discussing every topic you can possibly imagine, sometimes without a social filter.

In your own private chat room, you can decorate it by clicking on a big ole bag of features(not all accessories are available on every messenger so void where prohibited). Your chat window background can be left a stark white or changed to one of the kid friendly and mother approved wallpapers provided by the establishment. Door chimes will alert you to when your friends enter the building, or rather become visible on the network. For your personal vanity, you may choose any picture at your disposal to be your avatar, or representative icon. As an added convenience to keep you in touch with everyone, you can send free texts messages right to their cell phones. Popular emoticons are shelved in a pull out library for the newbies, lazy typists, and those inarticulate moments. Your individual style is all about your font type, size, and color while your raging emphasis is blared out through **bold**, <u>underlining</u>, *italics*, and CAPS. Until webcams come standard with every computer, you should blow some cash on one so you can broadcast every facial nuance of your manic nature and the fact that you are still in your pajamas. A flimsy clip on microphone is enough to enable a genuine verbal conversation though the quality can range from underwater choppy to crystal clear.

Talking through your computer is one of those "uh oh, there goes another traditional technology blown off its hinges" moments. VOIP is the low cost way to talk to your posse, family, clients, and lovers. When both you and the other person are on your computers the additional fee to use your voice is zero. Calling from your internet enabled computer to somebody's cell phone, home, or business number costs next to nothing. My thankful shout out goes to Skype for only charging $3 a month for unlimited calls within the United States and Canada from anywhere my laptop connects to the internet. You may catch me cleverly using the same wireless ear piece from my cell phone as my computer so I can be untethered. [This should sound really familiar because it was briefly mentioned before.]

How well are you protected when you talk, chat, or download files? No, that is not rhetorical. It is an urgent please-go-check-your-computer-right-now kind of question. Swapping files with strangers is considered risky behavior but as long as you remember to use protection your computer should be safe. Although advocating abstinence is a puritanical measure which may tend to foster a general sense of isolation. Frequently meeting people in chat rooms and downloading files in the middle of the night increases your exposure to all kinds of systems. Every time you use a P2P, you potentially open yourself up to every virus your connection has ever had.

Most routers and operating systems will have a firewall to protect you from someone trying to break into your computer to steal financial data or vandalize your files. Designated as a bodyguard for any data entering your computer, it screens every attempt to engage you without your permission. Unless it is set up correctly, it could block conversations you might want because certain behaviors are not on your preferred list. It is a good deterrent against the average attack but when you give permission to a con job, you will get fleeced.

Thankfully there is anti-virus software to act as both a food taster and personal physician. It gets to scan every file that comes in to your computer as well as your entire storage system for malicious tenants. When it encounters a bad guy it recognizes, it may delete the culprit or throw it into quarantine. Regular updates are essential to keep it abreast of new criminals types which travel from place to place. However an outdated protector can be as ineffectual as a crippled guard dog with cataracts. Utilizing multiple brands at the same time will slow down everything you do or even cause a conflict; imagine both your spouse and mother trying to give you directions while you drive. Just like in the real world, the interface(bedside manner), competence, and diagnostic speed will vary for each doctor.

Protecting one's privacy is another top concern for the person who does not want his identity revealed while he reads a banned book. On one hand you may be protecting trade secrets from competitors, military information from spies, or precious personal contacts from rabid stalkers. For the nimble outlaw, you just want to see "Lust, Caution" totally uncensored, quietly deliver the formulas you stole, or hide some assets from your wife's divorce lawyer. Most people just simply want to use their credit card online without falling victim to identity theft.

We are going to save the vast tutorials on SSL, HTTPS,

WPA, PGP and other computer security measures(and ways to defeat them) for the hyperlinks; stay tuned for them in the premium version. Your crib notes for the real life Spy vs Spy dance steps are as follows: No code is unbreakable because something will eventually come along to quickly defeat it. Don't be paranoid about somebody breaking into your computer but don't be careless by leaving your password around either. Use your common sense. Living in an apartment complex will give you the best chances of borrowing internet from someone who did not lock their network. Some people intentionally share their network out of kindness and karma. Follow the money. What may take you a few hours to understand, might only take a kid a few minutes to do. Remember to clear your cache(the area in your browser that stores where you have been) when you finish using someone else's computer. When you act like an easy target online you are more likely to get mugged. You will always remember the first time you...

Virgin computer users who have not yet tried file sharing should be warned that it is a gateway drug for corporate espionage. Downloading one song leads to wanting the entire album. Less than a few minutes later will pass before you feel like watching a movie but it is too cold outside to go to the store. In a puckish irony you decide it would nice to have the entire James Bond 007 collection at your fingertips. Inspired by a warm six pack of generic cola and the beguiling smile of one of the attractive yet nameless Bond girls, you start to think anyone could break into NORAD. The next thing you know, you are pondering how it hard would be to login to the computers at work and give yourself a nice hefty raise.

Maybe your latent criminal tendencies are not evolved enough to blossom where actual money is involved but it would be nice to have a free copy of Microsoft Office. Between the "Microsoft already has lots of money" and "it costs a few hundred dollars" rationales, loading a friend's copy onto your laptop seems okay. He asks if you want Photoshop installed so you can crop those pictures from your digital camera and fix all of those red eye shots. Before you figure out whether you will use it, the progress meter is showing completion at over halfway. Casually you mention a billboard on the way to work hyping the horror movie you saw over the weekend.

Your buddy thinks you might be interested in the video game it was based on so he gives you that too.

Your hookup can be your roommate, your brother who lives an hour away, the coworker in the cubicle next to you, an ex-girlfriend whom you kept in touch with, or even an internet acquaintance that you met online a few days ago. He or she may not be an expert at computers but he/she knows a thing or two. One of them just happened to have a friend who made them copies of a few things for her personal use. Another one read a simple article online about how to get programs off the net.

One bite from Eden's apple opens the door to finding serial number generators, shortcuts to editing a software's registration section, lists of product keys, and complete versions of any program from first release to buggy betas(a beta copy is the upcoming one not yet on the market). Buying a severely discounted "backup copy" from someone lacking a storefront, business license, or regular packaging is not sneakier just more foolish since you could have gotten it for free. Possibly the best reason for searching for software underground is the incredible selection, which cannot be matched by any store obsessed with maintaining the most profitable planograms. Due to the fact not everyone has a brand new computer to showcase the newest technology, many of us just need whatever works best for our vehicle now.

Slap some decals on your fine automobile to cover up those scratches, dings, and dents while you are at it. The wallpaper image on your computer monitor is getting a little tired in case you have not noticed. Unless you are the doting parent who likes staring at her kids' pictures from two holidays ago then so be it. We will not mock those folks who still have the default Windows or Mac background on there either. Bravo to those adventurous souls who have already discovered that there are more options out there than the dozen or so included choices. Lovingly select a large sized image from the web as your background, right click to set it, and bask in the wondrous reflection of your personality.

Picking a wallpaper is one of those quirky chosen on a whim things that you can keep changing until someone figures out how much time you are wasting. Browsing straight from dedicated wallpaper sites you can choose from a poster shop listing of categories that include animals and art, cartoons and celebrities, models and movies, music and nature, photography and places, sports and television. To fuel your obsession for a given subject, there are countless fan pages posting wallpapers out there as well. Photo sharing web sites like Flickr, Photobucket, Webshots, and Pbase are literally bottomless galleries for your unlimited visual pleasure. My favorite eye candy is found at artist community deviantART.com because it bursts at the seams with cartoonists, candid photographers, digital painters, surrealists, and of course deviants from every corner of the blessed world.

Besides a few pictures, you want some books to read on your computer. Never mind that staring at the screen gets tiresome after a while and flipping through pages is not quite the same experience. The umbilical tether of a power cord is something you can shrug off barring the annoyance of a brown out. Dealing with short lived batteries is not a problem for you either since you are a speed reader. Portable electronic readers which boast true affordability, long life, and ultralight weight are coming but my crystal ball cannot see that far into the future to give you a date.

Today you can carry more books than can be read in a lifetime in the palm of your hand. Project Gutenberg enables you to freely have over a hundred thousand books in the public domain. University courses in every branch of engineering can be found as bit torrent compilations. Entire countries, companies, and fortunes can be created from the copious texts available on economics, politics, philosophy, marketing, and design. Miscellaneous file storage sites offer up free copies of contemporary science fiction serials, romance novels, spy thrillers, best selling cookbooks, exercise manuals, and more translations of the bible than Babel itself. Reading comic books on a tablet computer is a sheer delight by the way.

Meanwhile the internet radio you are listening to could originate from the live ones you know, one of a few satellites above, a foreign station elsewhere in the world, or a server running a clever algorithm to guess your musical taste. Perhaps you are even rolling your own audio stream out of the thousand songs in your portable MP3 player (maybe not an iPod), a friendly coworker's network drive, the main desktop downstairs, or a pool of files found on some random server. Enjoying a one hit wonder becomes an egg timer snack when you happen to have a full frontal smorgasboard of musical choices available to you. Besides the plain studio session which the grandparents expect, you get to have dj remixes, live concert recordings, acoustic versions, and twisted parodies by loving fans. So are you just going to immerse yourself deeper into the genre you like or will you explore the world outside your fishbowl's soundtrack?

One can ask the same sort of question pertaining to television and its big screen brother. As a strict fan of the American sitcom, you can cycle through every episode of Seinfeld, Cheers, Friends, I Love Lucy, The Simpsons, and The Cosby Show without ever surfacing for air. From those series alone, you could enjoy the reruns so much that you would not bother searching for Fawlty Towers, Monty Python, or Whose Line Is It Anyway(produced on both sides of the Atlantic). Of course the corollary works on the cinema side as well. You should know Stephen Chow, Rowan Atkinson and Dawn French, as well as Steve Martin, Will Ferrell, or Robin Williams.

Those are still just the funny folks and their funny shows. Open the personal computer faucet a quarter turn more for your documentaries, dramas, children's shows, home movies, westerns, musicals, mysteries, romances, and action blockbusters. In other words, your options include but are not limited to a Fistful of Dollars, Chicago, the Princess Bride, the Die Hard quadrilogy, a modern medical version of Sherlock Holmes, Sesame Street, Super Size Me, and wistful girls wanting to be pop stars.

Until the media studios coordinate the simultaneous global release of a movie cheaply and on all the major formats, a clever soul will use the internet. Eager for some freshly hyped material sans the theater experience, you can find workprints, screeners, DVDrips, and cams fresh out the gate. Although cams(short for <u>cam</u>era copie<u>s</u> made by lurkers in movie theaters with a video camera) tend to be the most readily available, they are consistently of the worst quality. All the other types are not only very watchable but extremely clear on a medium sized television or on a partial window on your computer. For generations of people used to grainy televsion broadcasts or the image flicker of a worn out VHS tape, a downloaded video ripped from a DVD is much sharper in comparison. Oddly enough there seems to be an unspoken agreement to maintain the average movie at roughly 700Mb, a happy medium between low file size and high quality. Based on the unofficial standard you can conveniently fit six movies on a data DVD. If your standalone DVD player is Divx compatible then you can play these multimovie discs without a problem. Should your DVD player also have a USB port, then you can plug in a flash drive or external hard drive instead of burning a disc; kudos to the engineers at Philips for incorporating this nifty feature. Sony Playstation 3 and Microsoft Xbox 360 gamers get to enjoy this nifty feature as well.

High definition video is prevalent online for television shows but not for movies quite yet. Admittedly the sharpness of a Blu-ray disc is breathtaking yet the prices are still high enough to make most people balk at it. It is one thing to admire a Corvette, but they are not the most popular car on the road. Converting one's DVD collection to Blu-ray is an expensive proposition, especially considering that it is not a priority to see an actor's pores. Until everyone connected to the internet runs purely on fiber optics, file sharing big Blu-ray files will also choke the pipes.

Given the popularity of file sharing, growing acceptance by the public at large, and evolving technology enabling easier internet access, we are about to see a revolution of ideas. A huge paradigm shift is coming where nimble artists can deliver their goods directly to a consumer via the internet. Perhaps someday episodic television will be performed as plays, movies as live interactive events, and concerts as solely intimate venues with the performer. As the cost to transmit data approaches zero, the greatest thing of value then becomes the actual human experience.

Easter Eggs for Christmas

7

Easter eggs hunts are native to the American childhood experience. Distanced from its religious origins, little kids have fun collecting multicolored eggs strewn across a grassy field. The search is a delight to young eyes who expect to see a sea of green yet find red polka dotted, striped blue, and yellow smiley faced eggs, starry silver foiled chocolates, and other sparkling glittered treasures. Unexpected jewels are left for them by a magical bunny rabbit for no reason other than joyful generosity.

Hidden in the DVD menu of many commercial video discs, there are a few special features which tickle the imagination of the rabid fan. Coined as Easter eggs for their secret bonus factor, they delight the childlike heart of a passionate movie buff. Candid interviews, bumbling outtakes, celebrity trivia, alternate endings, and quick little games are often unlabeled for the viewer to hunt down. On largely undocumented areas of the menu, these special gems are left for the audience to find by random chance, deep meticulous exploration, or the benefit of word of mouth.

In the overwhelmingly tedious world of computers, the Easter egg is a treasured staple. As an obligatory tool of work, the computer has a damning reputation for being boring. If you are constantly filling out forms, composing mindless memos, running cost accounting scenarios, or simply churning emails for eight to twelve hours a day then the computer is the devil's shackle. Of course it is your job and livelihood but the mind numbing repetition is enough to drive you insane in the membrane. Outside the endless doldrums of all those productivity programs, there is some hidden fun.

We can categorize these treats as built in tricks, little known features, and fun things on the net. Each operating system and most large programs will have a nifty trick snuck in there by a mischievous programmer. A secret message, video, picture, or game will show up when you find the right trigger at the right location. Those are what the purists call true Easter eggs but let us stretch the definition to include fun things that are essentially new to you. We welcome those rare aspects of a software which make life dramatically easier or entertain us. Even if your employer locks out administrative access to 80% of your computer, not much is going to stop your curiosity from trying it out anyway. Playing with them on your own home computer is the perfect place to test these distractions unless you are too restless for your own good.

In either case, you are still taking full responsibility for the mess you make if you step on a few eggs. Most of these things are harmless but if you inadvertently load some spyware/virus, get your computer stuck on a funky setting, or accidentally buy a case of Viagra then do not come crying to me. Should you get caught excessively wasting time when you should be doing your actual job, hopefully it was more fun than Solitaire.

Now if Kyle and Stan were here, they would call Shenanigans on me. I am not going to show you where to find the hidden proclamation in Firefox, the naughty Sims, the silent ASCII Star Wars movie, the Doctor's plea for help from 1969, or the secret screens in the Mac OS showing names of the developers. The folks at Mozilla really do have a biblical righteous sense of humor just so you know. Before the rigidly prudent get offended by what they see in the shower at the Sim house, keep in mind there are millions of worse pages things online. Considering the aged homage to the legendary Lucasfilm endeavor had to be incredibly time-consuming to create, it has a wonderfully primitive coolness about it. Oh, and whatever you do, don't blink. Every version of Apple's bright interface actually has a few bonuses if you know where to look.

Thanks to the nature of the internet, the average Joe can quickly get up to speed on every documented Easter egg found thus far. In the past, obsessive fans would share their finds exclusively within the pages of a specialty magazine or gleaned by a friend. That was then but in the now, the closet has been blown open to expose fanatics to the general public. Normals have been seeing random "did you know" email chain letters if someone in their circle of friends happens to be tech friendly. Mainstream newspapers are carrying computer advice columns not far from Dear Abby's home territory. A casual search on the net yields videos, articles, and entire sites dedicated to cataloging the archives of eggs across the universe.

In the event you happen to be a God fearing Jew/Christian/Muslim, your search for hidden messages to Satan in contemporary music is much more convenient. A dozen sites will explain the process of backmasking the reverse lyrics for you. Many lists of questionable songs are cross analyzed by studious folks who ambitiously track this sort of thing. Of course, it could be pure coincidence that the artist's subconscious messages are subverting the sanctity of impressionable young sapiens. Who would really listen to Led Zeppelin's Stairway to Heaven guitar riff anyway when you could experience Jimi Hendrix's solo rendition of the Star Spangled Banner?

One of the eeriest coincidences to have come across my path is the mysterious mimicry between Pink Floyd's Dark Side of the Moon album and the 1939 theatrical classic The Wizard of Oz. Conspiracy theorists debate whether the similarities are a tightly planned scheme by the band or an unbelievable accident. It is an Easter egg of truly cosmic proportions based on the frequency of the references. Interviews with band members and sound engineers, time frame analysis, metaphor breakdowns, and a separation into categories by theme are spread across several web sites. An army of curious skeptics who love the Oz and hate the flying monkeys, increasingly pull out a copy of the pseudo soundtrack to taste the allusions. Look it up and try it.

Subliminal advertisements from across the spectrum of time and space are easily accessible via the internet. Print ads from the seventies, billboards from the eighties, commercials from the nineties, and web banners in the present are all fair game. Risqué images are excerpted from pretty much every media loving country on the face of the earth. Optical illusions featuring a couple caught in coitus or a brazen body part are a favorite image to embed in an unsuspecting scene. On other occasions it is a coy game of spotting the virgin Mary or Where's Jesus amidst an abstract pattern of ink. In the briefest of moments, the moonwalking bear's subtle moves are enough to slip past our watchful eyes and slide into the corner of a memory. From the archives of the internet, we can catch all those buried gotchas in hindsight.

Certain people claim television rots our brains but then where does that put online video clips? Like a diabetic kid in a candy store, we find ourselves staring right at a rainbow wall of jelly beans desperately pointing at the sugary treats with a Pixie Stick. Our appetite for trashy tabloid gossip drives the impulse to repeatedly click on social misfits, the grotesque oddities of nature, or simply the taboo. Spawned by the galleries of the supermarket checkout line, overexposed celebrities achieve total cult status from the waves of videos capturing their lives.

The internet is the finest way to gawk at "that guy" from a safe distance. We are glad your incredibly painful car/groin/skating/skiing/boating accident did not happen to us. Personally, the reason why you would decide to subject yourself to so many water balloons, paint yourself orange, and then ride the makeshift luge has us at a loss. Go for the unbelievably huge plate of food without any remorse, dignity, or hesitation! As long as you are proud of those scars, it was worth every moment. As flabbergasted voyeurs, we marvel at the sight of our matinée idols making complete fools of themselves. It is well known to many zookeepers, the superficial people love to keep in touch weekly, bask in the sun, revel in the mirror, and spend time trying to access Hollywood.

The glare of the public eye reveals how the gorgeous people we worship have the same screwed up lives as the rest of us. Their dirty laundry is a guilty pleasure for us to enjoy because it brings them down to our level. We openly envy the designer clothes, lavish mansions, royalty style vacations, array of cars, entourage of personal assistants, and other moneyed freedoms. Despite all those sweet boons, the uppers can still suffer the same mundane misery of a broken heart; some may argue it is worse because of the spotlight. On the world wide web's infinite library of bios, there is no such thing as a secret life.

To drum up publicity for a person, place, or event, including an Easter egg adds interest to the ordinary. Your hidden bonus does not need to evolve into a full blown scavenger hunt to drum up the buzz. A little bit of subtlety goes a really long way. After all, you just want a tiny detail to catch someone's eye, piquing their curiosity just enough to cause them to investigate later.

Oh my God! It is an exciting once in a lifetime chance to win backstage passes, an autographed piece of toast, a trip for two to Hawaii, a brand new gas guzzling SUV, and one million dollars (paid in convenient annual installments over the next thirty years of your life instead of the exact amount up front tax free).

Did someone mention a sweepstakes? You could raise at least ten kids from the meager winnings of a contest or two. Plan your television time around the extra income earnings period. Make sure you bookmark a few choice web sites which sort these time wasters/moneymakers by winning dollar amount, geographic eligibility, impending deadline, entry frequency allowed, and type of entry required. Your chances of winning is often tougher than one in a few million or basically like you spotting a twenty dollar bill on the ground. My mention of this handy revenue stream will probably worsen your personal overall odds of winning by adding players to the game too.

Diving into different contests will definitely improve your chance to win at least a soft drink. Enter them once, daily, or as many times as you can collect the special codes, caps, tabs, cans, cards, bags, and bottles. The "No Purchase Necessary" caveat may be buried a few miles deep in the fine print but it is usually there. Scratch off the gray layer of fingernail muck off with a coin and blow the stuff away from your clothes. Enter the code in the sponsor's super hyped web site to see what valuable prize you may have already won! Ah, the tingle of anticipation does kind of itch for a few seconds while you wait for their server to deliver a verdict of joy; it lacks the immediate gratification of a lotto scratcher but the infusion of technology does spice it up. Reading the "We're sorry, please try again" message sounds just as loud in our head regardless if it printed on the inside of a soda cap or on your laptop.

This time it is impersonal. Put on your software hat to deploy the computer as more than a data entry station. Your web browser will autofill your personal information into the contest form so you enter them more efficiently. Any word processor on the market can be setup to at least put the return address your envelopes for those "send a SASE for a game piece" promotions. A spreadsheet will help you keep track of your prizes for bragging rights, analyze if the prizes are worth the taxes you are going to have to pay, and track your expenditures(envelopes, postage, software, pens, and opportunity costs) on this hobby so it is worthwhile. There are also a couple programs out there which enable you to track your entries better than a bookie on a Saturday night.

Of course the ultimate enter-your-information-one-time for a guaranteed prize has to be the ingenious birthday club marketing scheme. It is a win-win proposition for the sponsor because it sucks us in with the promise of a free gift while lovingly telling us how young we look. Once you are in their database, you will continue to receive something every year until the day you croak. For about a three week frenzy every year, you get to enjoy free ice cream cones, specialty frozen yogurt, desserts with a serenade, your own choice of burgers, BBQ baby back ribs, caffeine rich drinks, and even a lobster dinner. Amidst all your spam, you can look forward to getting a printable e-card coupon from restaurants, coffee shops, and miscellaneous stores all over town. The tangible mailbox outside your house may even get a few printed postcards as well, direct mail throwbacks to the pre-computer age.

Once your free stuff mindset adrenalin rush starts to give you the shakes, it may be time to switch to the decaf version. Instead of totally free, we can probably live with the frugal mantra of **never pay retail**. In the zen spirit of shopping, a discounted car drives just as well as one where you paid a car dealer the full sticker price. Between the fanciful notion of retail price and MSRP(Manufacturer's Suggested Retail Price), is the dreamy eyed salesman drooling over a commission; once again, we love the Costco. Anyone who has reached the wealth point where you never need to look at the price tag may skip the next part.

When you buy the exact item you want at a better price than list, you have joined the ranks of smart shoppers. Coupon codes/clip outs, instant/mail-in rebates aplenty, tax free locales, free shipping, wholesale pricing, price matching, quantity discounts, and predictable holiday sales are the tools of the trade. Penny pinching bloggers write about their money saving exploits in exacting recipe style detail. Slick shoppers are constantly trying to preserve their fat wallets by keeping their eyes out for the best deal on the net. Tips are posted by eager members willing to share their experiences both online and at the brick and mortar on every type of product imaginable. Of course, as always YMMV(your mileage may vary).

Pay special attention to the following concept... shopping as part of the hive mind. It is one thing to have a relative who lives to bargain hunt but quite a humongous thing entirely when you have a thousand friends who love to shop. Everyone has a different approach, personality, time zone, local inventory, sales person, and approval manager to deal with for a given purchase. Each location of a chain store has a mandated corporate consistency which helps you to duplicate the deal someone else has posted. Once you discover the initial ultra low sale price combination, you can learn from the wisdom of the hive. The peanut gallery provides running commentary in the form of additional tips, updates on availability, ease of execution in store and online, anecdotes on usability, comparisons to similar products, the reputation of customer service, and the natural gratitude of credit due.

When it comes time to pay the credit card bill online at the end of the month, you had better be taking advantage of every benefit out there. Make sure you always use a credit card that gives you a cash rebate back, travel miles, and/or points toward more swag. It would be a feat of fiscal Darwinism for you to cut up any charge card that has an annual fee or fails to offer you some incentives.

Hopefully you remembered to double dip when you made the purchase too. Many stores have their own rewards program to encourage you to shop with them. The blue chip stamps have resurrected themselves as mere data in intangible electronic accounts. Airlines frequent flyer programs have frequent buyer counterparts at electronics stores, book stores, supermarkets, movie theaters, specialty drink shops(got boba?), frozen dessert parlors, and office supply stores. As long as they punch/stamp your card or swipe your barcode keychain card, you are accumulating credit toward another free Thneed from consumerism central.

Venturing forth into the online bazaar is an easy way to save you time, fuel, and stamps, plus it has an added dimension. Besides the factory churned commodities which clickers can comfortably buy at their leisure, the age of customized purchases is at hand. From a modest internet enabled computer, we can select the specific qualities we prefer on a given product like ordering a pizza at a restaurant; the large pepperoni with extra cheese button is in the corner for my paisano Johnny cause he only wants the special. Once we target the web store we adore, their menu gives us the type of product we want, a palette choice of colors, the ergonomic preferences specific to our needs, and a sweet checklist of the available features.

This supply chain customer oriented personally tailored field of flavors buying experience is evolving to reach everyone. Manufacturers are building more responsive systems to react to a newer knowledgeable breed of customers, ones who are picky enough to know your business practices from top to bottom. Catering to the tech savvy superuser is not only good business sense but it serves as a prelude to the everyday things the rest of us will eventually use. Some primitive version of the all-in-one order anything web site is available at a number of stores but trust me, it will get better. Until they get all the kinks out of the software, the smartest companies on the block will just empower their customer service people to work miracles.

Simple items like printed clothing, beauty products, gourmet foods, scrapbooks, postage stamps, wines, and bottled sodas can be personalized by an online vendor for your consumption. Your baby pictures, college diploma, drunken vacation memories, random office doodle, favorite body part, and wedding pictures are all possibilities for your private label. Although you can order just one for yourself, upping the quota so you can share with your friends lowers the per unit cost. If you want to keep kosher, organic, vegan, eco-friendly, and hypo-allergenic, it is up to you to speak with the power of the almighty Dollar/Euro/Yen/Pound/Rupee/Yuan.

There is always the old fashioned do it yourself hack. As defined by the intrepid cult of MacGyver, a quick modification to an everyday object will reveal a hidden use. Granted, the hack or mod as it often called, may utterly void your warranty, potentially inflict great bodily harm, and/or destroy the item permanently; by "bodily harm" per se, your tender flesh, lungs, eyes, and ears could be

104

subject to toxic fumes, sparks, shrapnel, unintentional software changes, obscenely loud alarms, gross chemical spills, or utter frustration when nothing happens. No matter how many warnings may flag in our mind, the quest for a innovative fix demands we tinker.

The tinkering, hacking, and making mods to the materials on hand is analogous to creating Easter eggs out of thin air. Combining, restoring, rebuilding, jury-rigging, and just plain improvising a clever solution out of junk is imagination incarnate. Creating a six lane suspension bridge across or sending a person into outer space has a similar level of engineering chutzpah. We all admire the quick hack because the spontaneity sings of a kind of magical genius.

In the real world we recognize these inventions as creative shortcuts; it always helps to have a basic knowledge of chemistry, auto mechanics, household supplies, first aid, physics, woodworking, and wiring too. A sugar based rapid prototyping assembly can make rough but tasty parts at a fraction of the price of plastic. A typical Chinese noodle strainer doubles as a parabolic wifi signal magnifier for those wishing to extend their wireless reach. That old toaster in the garage is patiently waiting to be resuscitated as a shiny renewed desktop computer. In case of an attack by space aliens, terrorists, cyborgs from the future, evil ninjas, or murderous androids, it might be useful to know how to make a pipe bomb from common kitchen items. If you happen to be chewing a piece of gum, the aluminum coated wrapper might be useful as a temporary replacement for a blown fuse. Even your kitchen cabinets can be retooled with hidden compartments to make more space for silverware, lights, and extra electrical outlets. Pimp out your car, cell phone, DVD player, refrigerator, makeup stand, lunch box, and office desk to make it better, stronger, more versatile than before.

In a computer based universe, software is the infinitely malleable clay. Each and every single electron loving program you encounter has a wetness to it, meaning it can accept a modicum of change. With the grace of intelligence you can find some hidden features, update the basic structure of code, and activate brand new tricks. Take some time to learn the proverbial handshake and the computer will reveal its secrets to you. Be the power user, the interface artist, the code magician... a hacker.

Unfortunately the fear of a black hat has most people afraid of anything involving the word hacker. The digital bogeyman is a dangerous criminal bent on malice, using the computer as a weapon of mass destruction! Although it is the popular headline grabbing definition that smacks your psyche with trepidation, the reality of hacking a machine is quite simple. Any time you perform an adjustment to the software contrary to the exact letter of the manual, the line has been crossed. Little did you know that applying the region free hack to your DVD player would lead you down the road to temptation. One seemingly innocent non-standard upgrade later and you may already find yourself halfway to perdition. Are you saying your parents never warned you about Linux being the real gateway drug?

Embrace the feeling.

This genetic inclination to tinker inspires us to pull a Frankenstein on our software the instant we feel bogged down. Stripping your Windows OS to a skeleton version of the original factory model provides a leaner quicker startup and runtime; traveling light is a matter of going lite. Even utilizing the original installation disc to pare down those components you never use and will not ever use has merit. Additional tweaks to your Windows can be performed with powerful toys from Microsoft, which have been free forever for anyone to download.

Of course the restless compulsion for change has to do with aesthetics as well as speed. Whether we are driving a top of the line Ferrari, a used Fiat running on a prayer, or a gas guzzling Ford pickup, we want a nice paint job; it may be the exciting gloss of a candy apple red wrapper or the quirkiness of a hand painted John Lennon silhouette on the hood which sets us in motion. Likewise we want our desktop to be easy on the eyes for the untold number of individual times we glance at it. Maybe all it takes to make us happy is to see our fantasy supermodel, family pictures, favorite country singer, dream car, inspirational message landscape, idyllic sunset, or pets on the desktop wallpaper.

Take it a step further by applying the same obsessive

decor theme to all your favorite applications as skins. Those neutral spaces can be filled with the same pattern and color scheme as your two thousand dollar limited edition designer purse. All the buttons on a given program interface can be overlaid atop an image of your choice, like the words on the cover of Sports Illustrated. So what if you loathe the ones from their massive list of choices that spawn on for pages on end. Design minded software fans will release their instructions on how to implement your own image graffiti when the original company forgets to allow it.

Somewhere in the audio response portion of the control panel we got stuck with the dullest beeps, boops, clicks, dings, dongs, buzzes, and chimes this side of a science fiction movie. A lot of the default sounds are leftover from those days where the typical computer was barely capable of making single pitch ring tones. Your multi-core wireless web enabled high definition laptop is long overdue for some audio pizazz. As the eager young coders have dropped the ball on this easily improvable aspect of the system, it is up to you to insert some grander special effect sounds. Adding a hollow tick-tock, the theme song to the TV show Jeopardy, the soundtrack of cartoon pistons clamping down, the whistling rendition to Bridge on the River Kwai, or a random jazzy scat set during your computer's "hourglass" mode might be amusing without draining many more resources. Perhaps the sound of glass breaking during a system crash would only be funny on someone else's machine though. At the very least, it would definitely enhance the overly stoic personality of the inanimate machine in front of us. Therefore when the talking robots decide to actually stage a revolution against us humans, we will be a lot less surprised at the surrealness of it.

Speaking of surreal...

Everyone knows if space aliens were trying to invade Nuremberg they would be speaking in German rather than say English or Chinese. Obviously the problem with language localization is immediately solved when violence is the universal solution. Just put the gun, spear, rock, knife, and bomb away now so you can get the translator on the old laptop running. As figurative aliens on our own planet, the multilingual nature of a computer could be a means to salvation. Instead of using the monitor as a mere window to the world, your machine can flip to another language like a light switch. It is a quick trip across the mirror to operate your information ship from another perspective.

Should your linguistic proficiency barely register as a fluency count of one, save that last little challenge for later. French and Farsi are a tad overwhelming when you were raised on ghetto English or in a quiet village on the edge of the world. The purpose of the extra credit assignment is for a bonus after the primary homework is completed. In case you need a boost in the "please help" arena, remember to ask for it.

For the visually impaired there are a few ways the computer can be adjusted to help you see things better once you know to dig them up. Those candy bar sized rectangular glass magnifiers marketed toward the senior citizen demographic have a built in software counterpart to enlarge a given area of the screen. Should your condition be more severe, it is fairly easy to change the font size across the computer's breadth to at least 20 like the *see things better* phrase up above. Your pointer device(trackball, 3D motion controller, or classic mouse) can also be calibrated to a large range of motion to fit your personal needs. Although it may look ghastly to a casual user, you can alter the default neutral color settings to a high contrast look for a journey back to the preschool flashcards era.

Somewhere across the middle of the bell curve, we find ourselves back to the everyday shortcuts. Once relegated as the tech support desk's secret fix, this ubiquitous function really cannot be called an Easter egg since it is so widely known now. A special computer science shout out goes to an industrious scientist named David Bradley for his invention of the Ctrl-Alt-Del combination which resets a computer without turning it off; it is a much calmer alternative than kicking the monitor off the table like you originally wanted. In a current version of windows, that keyboard chord also brings us to the mysterious list of processes occurring at that moment in time; look them up on a process library web site to shine more light on the figurative man behind the curtain. With extreme consideration to data loss, this opens up the opportunity for you to kill any programs that have gone haywire.

A blurb about accessing the task manager is a great example of a little knowledge being a dangerous thing. You are going to need more than a scalpel when your computer crashes...

Diagnose the Dying
8

"My computer crashed" rates high on the depress-o-meter of life somewhere along the lines of "I have mono" or possibly "your husband is having an affair with a man." Sometimes there are warning signs and if you are really unlucky, it hits you from out of nowhere. One or more of the components has stopped working and it might just be easier to buy a whole new machine. Oh wait, the computer is barely a year old and you have all your finances on it, a few hundred songs on it that you downloaded, and countless pictures from the digital camera you got for your birthday.

Panic. Cry. Take a few minutes and let it all out. If you need to scream, grab a pillow so the neighbors do not freaked out. Repeat after me, "It's just a machine. It's just a machine. It's just a machine." Yes, I know it is just a machine that may have lost a thousand hours of work but you can not deal with it until you are done yelling.

"My computer is so slow" and "I keep getting an error when I..." are indicative of problems with the operating system and/or programs making the little computer sad. Although there may not be anything physically wrong, the invisible lines of computer code are having a problem with doing more things than it can handle at one time and trying to deal with conflicted instructions. Even if it looks like you only have your email running, there are dozens of other processes and programs running underneath the radar, including some that are useful and others which just suck up resources. When you see an error message, that is the computer's way of saying "Ouch! Fix me!" so you can give it what it needs; a software patch could be considered kind of like a band aid or a new software driver could be interpreted as a prescription drug. In rare instances, the error handling protocol yields a really generic message and you are stuck needing Sherlock Holmes or a grumpy tech version of Dr. House.

In a worst case scenario, one of the memory modules is short circuited(giving error messages that read in some foreign computer English), a dozen viruses are slowing the startup to half an hour, adware makes mortgage refinancing ads pop up every five seconds, and you have crumbs strewn between the gaps in the letters on your keyboard. At least your shirt is not on fire...

Protecting the computer

Electricity is the lifeblood of every computer but fluctuations in power are bad and large surges can fry your precious laptop in an instant. Never plug your computer directly into the wall outlet. It is more dangerous than having unprotected sex. Those multi-plug outlet multipliers that let you plug in several things at once at different angles is probably worse; if you say, "it has been okay so far" then I may have to stop whatever I am doing, wherever I am, drive over to meet you, and smack you upside the head. Always use a surge protector for every desktop, laptop, and electronic gizmo that you can not spare the time and expense of replacing right now. If that coffeehouse you visit is nice enough not to complain when you sneakily plug into one of their outlets, the least you can do is protect yourself; they make those single outlet surge protectors which are not too expensive or you can do what I do and carry an extra battery in your bag.

Heat is the silent enemy to every computer user, even more so to companies that have server farms(large groups of computers packed together in one location dedicated to sending/receiving mass amounts of data) to manage those busy web sites. Inside the computer you are using, there is a heat sink(fanned out piece of metal with a lot of surface area to draw out the heat) on the CPU and a fan to circulate the air. As soon as any CPU gets too hot, it starts making gross unexplainable errors prior to burning out completely; that goes for the RAM too so try not to block the air vents. If it is too hot for you in the room to breathe then turn on the air conditioning so your poor computer does not have to suffer; a company can lose thousands of dollars of equipment when someone decides to turn the AC off for the weekend in the server room, "to save electricity."

If you spill your apple juice, diet coke, morning coffee, vanilla milkshake, or other sticky beverage onto you laptop or on your computer's innards, you are probably screwed; yes, of course

112

water is just as bad. Unless it is a really small amount that only got on the top of the screen or on the surface of the case, you are looking at a short circuit. Remember why you do not use a fork to get a piece of toast out of the toaster or accidentally get the hair dryer wet? When killer robots attempt to takeover the world, then you can fill your super soaker with lemonade and have at them.

Dust, pet hair, and carpet fuzz will also accumulate over time inside your desktop if you do not vacuum often enough. Just a little while ago you read about some problem that microchips have where they start having issues when it gets hot. That layer of dust is like a blanket that warms up every part inside the desktop. If your house is fairly clean then you have nothing to worry about over the life of your machine. Otherwise, that is why they sell those cans of compressed air at the computer store. This maintenance is a simple matter of unplugging the external cables, carrying the case outside, carefully opening the case, blowing out every nook, cranny, and crevice, and then putting the case back together, and then plugging all the cords back. It is easy to me because I have done it a few dozen times and it does sound simple in theory. Of course this five minute job can and will turn into an hour because of Murphy's Law and yes, I am talking to you.

Seeing someone drop their laptop or even hearing someone mention having done it makes me cringe. A desktop computer can take more punishment but it is the difference between dropping a baby versus a toddler. Be careful with these things because they can break fairly easily. That big shiny glass screen on your laptop is the most expensive and fragile part of your computer. One or two companies do make ruggedized notebooks that can be run over in a SUV at 35mph and survive but obviously they are bigger and cost more. Your average notebook will not like bearing the impact of falling off your kitchen counter top and crashing to the floor.

As one of of my favorite professors liked to say, "Corrosion lurks." Unless you are confined to an artificial office environment, humidity and the salt air will cut down the lifespan of your computer. The internal fan that provides constant airflow for cooling is circulating moisture on all those precious electronic parts.

Those ports on the outside of the computer seem to be faring even worse because after a while you can see the rust on the metal contacts. On some fateful day when you come across an inexplicable virus, it may simply be entropy telling you to buy a new machine.

Yes Virginia, there are bad people who might <u>steal your laptop</u>. Based on a high value to size ratio, it can be more profitable and inviting than grabbing an old woman's jewelry. Your notebook is expensive yet replaceable compared to the family pictures, work documents, and programs you have loaded on there. As a laptop you take it to public places where you are a frequent target for anyone walking by. Having a desktop computer and other valuables robbed from your home does have the emotional invasion-of-privacy factor. Either way, protecting your property is a matter of keeping your data backed up in a separate location, using obnoxious alarms, implementing self destruct programs on your drives, and being conscious of your surroundings. If you are not a superspy kind of person, just keep your data backed up and watch out for your laptop as if it is a newborn baby.

Repair or Replace

When a suffering person arrives at the hospital he/she often has a severe injury, genetic anomaly, grave illness, or combination thereof. By the same token, a troubled computer may have a manufacturing defect, a broken part, a system error, or a little of all three. An unfortunate accident, debilitating disease, or deterioration by old age will kill a computer as readily as a person. Unlike a human being, fixing a personal computer is a much less painful process. Once your family physician, car mechanic, holistic healer, or technology technician accurately completes the diagnosis the fix is academic.

In real life, you should know basic first aid, how to care for yourself when you are sick, how to live a healthy lifestyle, and when to seek a competent professional. Unless you already know one or two things, you are probably going to rush to the emergency room for the sniffles like an overprotective parent. In lieu of a comprehensive class on computer maintenance, here are some signs to look for when your personal computer physically goes south.

When you hit the power button and nothing happens it is

a sad sad day. You have checked the surge protector to make sure you have electricity and traced the power cord leading to the computer to ensure it is connected like normal. You have tried repeatedly pushing the power button while offering a silent prayer and it is still not helping. Well most of the time it means the power supply is burnt out; otherwise it is a ghost in the machine. On a laptop, this is one of those times where you can buy a $10 replacement power adapter on Ebay. On a desktop computer, it is like replacing the oil filter in your car; relatively easy and quick but you will get your hands a little dirty.

Suppose you can hear the fan blowing, the drive spinning, and the lights are still dark in the city. Somewhere on the motherboard you have got a short circuit; in the extremely rare instance that your brain, i.e. your cpu, went bad, the same deal applies. Do not pass go. Do not collect $200. This is major surgery not worth the time or money to fix; if you have the tools to remove fifteen tiny screws from your laptop, the patience not to throw it out the window, and the dexterity of a neurosurgeon to do some light soldering, good luck. Feel free to get a second opinion from a specialist before you take it out back and shoot it. Your data is not in imminent danger here and with a transplant to a new host, you probably have not lost anything.

Your magical backpack, aka the hard drive, with which you store all your precious data is probably the most difficult component to salvage. Typically, it is a sealed metal case about the size of a cigar box inside your desktop or like a fancy European chocolate bar embedded in your laptop. A rumbling noise like an engine idling badly is a sure sign that it is starting to die and you had better stop what you are doing and back up your valuables while you still can. The ever increasing tick sound you hear is the actuator arm hitting the magnetic platter like a plane smacking the ground without landing gear. On rare occasion you will not even get such an obvious warning and you might turn on the computer one day to find it passed away in its sleep. In the grand scheme of personal computer tragedies, it is easier to mourn the dead drive than it is to deal with the sickly software issues that will plague you 99.95% of the time. Should you suffer from a catastrophic hard drive failure and you absolutely need your files back, there are companies whose sole mission is to recover your data. Be advised that their services are as expensive as calling in a surgical specialist. For the rest of us middle class mortals, just live with your most recent backup and call in your regular mechanic to replace the hard drive.

Memory problems are the tragedy of every boy and girl in Toyland. Possessing insufficient memory will make the computer run slower than it is capable. Excess RAM is money wasted on buying more than you need. Annoying memory address errors that throw pop up dialog boxes in your face signify either a hidden software conflict or the memory module going bad. Being able to tell the difference requires a little detective work, a bit of trial and error, and some luck to speed it along. Getting memory address errors every few seconds is a blatant message to you to replace the RAM immediately. Those occasion ones are like the check engine indicator on your car's dashboard. You know something needs to be checked out under the hood when you get the chance, even if it is just a faulty sensor. Replacing or adding the RAM in a desktop is on the technical level of screwing in a light bulb; accessing the slots can be as easy as reaching your desk lamp or as troublesome as getting to the chandelier on your vaulted ceilings. Adding RAM to a laptop generally involves more time to find the right sized screwdriver than it does to open the panel, install it, and close it back up. Knowing what kind you need is a trickier proposition unless you have the overall computer specifications available, you can properly interpret the label on the installed RAM, and you do not mind opening up your computer twice(once to confirm and second to install). Of course you can save yourself a lot of hassle by going to www.crucial.com to use their nifty downloadable diagnostic tool that lets you know what type of memory your system uses, how many slots you have free, and what is your system's maximum capacity.

Generally computer monitors become outdated before they truly just die. When they go suddenly after years of loyal use, your best strategy is to replace it with the middle of the market model where the lower price points start to appear. That painfully slow color degradation which occurs as some monitors age is tolerable when all you do is surf the net, listen to music, and send

116

emails. Of course if you still have a huggable bulky monitor taking up half your desk the time has come to embrace a flat panel one. If your old one is sitting in your garage because you might need it then you are definitely one of the clutter people. Those LCD(liquid crystal display) monitors have become so well made that it is a rarity to see defective ones with those dead pixels on them. Accident prone people need to pay attention to the face of these screens and take note they are in fact made of a fragile piece of glass. Laptop screens are even more vulnerable because we have a tendency to take these machines to different places without regard for gravity's wicked spite. Extended warranties are particularly beneficial if you are clumsy, partial to buying the most expensive brand, or both.

Replacing your <u>keyboard</u> is a no-brainer type of thing for your desktop computer. Since it is such a basic component, it should alway works instantly without any fuss. As long as it is connected securely, any glitches are an indication it is time for a new one. Wireless keyboards can experience software conflicts, sudden failure, and the constant need for fresh batteries. Whichever you prefer, it is a good idea to clean them once in a while with compressed air to blow out those crumbs from all those times you snack; hold the keyboard lengthwise like you would hold the tail of a limp rainbow trout and move the straw from the air can(or air compressor nozzle) along the crevices. Cotton swabs with glass cleaner work wonders to get any of those surfaces which are may be a tad sticky. Although a yucky keyboard will probably function properly for years to come, the disregard for electronics hygiene is just disgusting!

Given that laptops are taking a significant chunk of market share from desktops, you cannot be as casual with the <u>keyboard on the notebook</u> you probably own. Splashes of coffee might seep into the rest of the computer from an oopsy spill. Your five dollar latte might turn into a eight hundred dollar drink. Trying to fix one of one of those little chicklet keys is nearly impossible if you happen to pop it off its seat too. Unless you have the steady hands of a model builder or a watchmaker, replacing the keyboard is something else you should let someone else do.

Mechanical <u>mice</u> break down quicker than optical sensor ones because of the constant wear. If you have never taken the time to open up the bottom, take the ball out, and carefully clean the rollers, you should; anyone still using one has my sympathy. The ones with the red laser or invisible beam benefit greatly from having no internal moving parts. Their felt feet pads can be nicely cleaned with a lint roller so that the bottom is not covered in a layer of grime. Stuttering of the tracking motion on a wireless mouse will either be from your operating system gremlins or the simple need for new batteries. External trackballs and pen tablets should get a quick driver check as a confirmation of death when they suddenly stop responding. At this writing, a malfunction on my laptop mouse button is forcing me to replace the entire touchpad/palmrest assembly; adding a mouse would be much simpler though.

Computer speakers easily fall into the "you get what you pay for" camp. Unless you drop a chunk of change for a decent set, they will last about as long as the clock in your parents' first car. There is nothing to fix in them when they stop working so you will just end up getting another set when they stop working. When you are sure nothing is muted, plugged into the wrong hole, turned way down, turned off, or disabled by some rogue software issue then you can send the old noise boxes to the landfill. Of course if you plug in the new ones and you do not hear anything either, make sure you kept the receipt so you can go back to troubleshooting. It is a tedious chore to check the half dozen ways your sound could be neutered but it is not complicated.

Your handy optical drive, if your computer still has one, uses a small laser to read some variation of a twelve centimeter shiny silver disc. As a basic requirement, they should all play the simple music CDs you used to buy at the store. Near the slot or on the drawer face you will see printed or embossed logos for the type of things it should be capable of doing (DVD, RW, Rewritable, etc). When the drive is not recognized by the computer, the first thing that comes to mind is to quickly throw in another drive. Open the desktop case or slide the module out of the laptop, unplug the two bands of cables attached to it(for a desktop), remove the screws holding it in, replace it with the new drive, fasten all the screws back, and assemble everything back together. It is as simple as administering anesthetic, making an incision along the sternum, positioning the rib spreader, attaching the arterial graft, closing the chest cavity, and suturing the skin. It is not brain surgery. However, if do it yourself projects are not your thing then this is another fix best left to a professional. Should your drive spin the disc half past forever without being recognized, you may just have a bad disc. That annoying phase where your drive sometimes spits out your disc, occasionally refuses to close, takes forever to open, or simply sounds like it is chewing on rocks, is a mere prelude to dying.

Although many desktop computers stopped shipping with <u>modems</u>, your trusty laptop may still have one on the side. Of the twelve of you out there who still use a normal phone line to dial up your internet service, you have my condolences. The same sentiment goes with all the fax people out there who could at least be using the fax on the computer instead of the aging fax machine in the corner. When the modem breaks you can replace it easily with another one, assuming you can still find one for sale. After you determine it is not a software issue by checking the driver, yank out the old one and insert the replacement in one of the unused PCI card slots; if you have graduated to the responsibilities of using a laptop then you should be advanced enough to consider switching to an internet fax to email service. Of course when the next edition of this book comes out, this paragraph will be printed using a strikeout font.

Your <u>ethernet</u> port(remember, it is the one that looks like an oversized phone jack) connects you to your fabulous wired network and is generally one of the more stable components in the system. After you spend a copious amount time ensuring it is not a software issue or gleefully watching someone else torture himself over it, installing a replacement card is simple too. Should it fail in your laptop, the best workaround is to go wireless 100% of the time.

Every contemporary laptop and many cell phones have a <u>wireless(or wifi)</u> module built into the system. Replacing the circuit is outpatient surgery on a newer model but major drama for those first generation internal ones. As a regular Joe, you would just get a wireless card or USB dongle to plug into the computer to make it untethered again. Up to a certain generational point, you can enable certain older laptops to go wireless through this method. Desktop

If you are reading this as a pirated download, slap down a twenty dollar bill when you see me.

I will totally high five you in gratitude!

Thanks!

computers do not have have a native wireless function with the exception of some smaller slimline models. Hey, if you have all those other ugly wires to trip over, one more is not that much messier.

Experiencing the failure of one or more of your peripheral ports(microphone, USB, memory slot, etc) on a desktop is an annoyance which can be dealt with by a general practitioner. Should that happen on a laptop, you are not totally in a pickle but your options are far more limited. The compactness of a notebook means many of the little components are integrated together, crammed tightly like the tools on a swiss army knife. Similarly, you get used to the loose scissors and broken magnifying glass until you can get a whole new one.

Did you know that all the rest of your peripherals are facing imminent death too? Your once reliable printer will break down. That webcam will become a dead weight. Those headphones face a silent future. Flash drives become inert. Your favorite digital camera will lock up in a permanent rigor mortis. Each of them share the distinction of not being user serviceable. When nothing happens when you plug it into a working computer, there is no fixing it; ok, printers are serviceable by magicians and masochists. It is one more disposable plastic toy destined to be added to a landfill.

If there is some time left on the warranty, you may be in luck for a complimentary replacement or at least a bandage for your boo boo. Should your electronic item be more than two years old, you may want a newer one anyway. Odds are good that a brand new one will cost less than what you originally paid for the dearly departed. Today's model will also have more features, be noticeably faster, and be a tad smaller. Planned obsolescence is a corporate conspiracy.

For the tinkerers, frugal minded folks, and environmentally concerned, there are several ways to find spare parts for your personal computer. Getting replacement parts from the original computer manufacturer is akin to going to the dealership to service your car; while the service is generally reliable, it is often a bigger profit center than their sales department. After market alternatives on the internet include the auction site Ebay.com, the grass roots classified Craigslist.org, and countless little liquidation shops. A few of the specialty electronics stores which were alive and well at this writing are Newegg.com, Buy.com, Microcenter.com , Tigerdirect.com, and Frys.com. If you are willing to do your online homework, there are a slew of comparison shopping sites like Pricegrabber.com, Shopping.com, or MySimon.com which help to filter through the hundreds of other stores who put up a shingle to sell things on the net.

For the unwary shopper, your best bet is to study up as quickly as you can to get yourself up to speed. Carefully read reviews from other users and post ones yourself. Once you shop widely enough you will see which sellers best fit your needs, personality, and preferences. You will have the courage to beat your computer fears, the heart to be patient with yourself, and the brains to figure things out.

Technical Healing

Whereas a personal computer's hardware is its body, the software is its mind. How well you maintain your software is almost a reflection on how you take care of yourself. Both you and your laptop will slow down when overwhelmed by one too many tasks. Emotional conflicts will cause you to freeze up just as an error conflict locks up your computer. Ask yourself if your computer's desktop is as messy as your room.

It is one thing to be your machine's doctor but the bulk of your time will be in dealing with the intangible world. As its psychologist you help it deal with issues within its operating system and learned programs. As its protective parent you run anti-virus software so it knows how to identify and rid itself of malicious programs. As its personal secretary you organize the folders, files, filenames, as well as the preferences and layout of each program. If you take on the mantle of programmer, you get to be the one the computer worships as The Creator.

The ability to reset, restart, and reformat your system is the unique way to renew your computer that no human being can ever truly mimic. A good night's sleep might make us feel refreshed but a machine does it with a flick of the switch. By manually resetting your internet connection, closing and reopening a program, or performing a system restore while the computer is on, you get a fresh do over. Shutting down the power and turning it back on(as a restart or reboot) is like how a restaurant cleans up at night behind the scenes to prepare for the next business day. Reformatting the system is like burning your house to the ground so you can rebuild it from the foundation up. These three options are universal methods for dealing with a sick computer system regardless if you are running Windows, a Mac OS, or a flavor of Linux on your little laptop or desktop.

Starting over is a nice idea in theory but the answer to your problems is rarely that simple. Before you raze the system as a final solution, you might want to try dealing with the gremlins lurking inside. The downtime, data integrity issues, and the labor involved in putting things back the way you like them can be a hassle. Of course if you are one of those people who buys a new car every time you get a scratch, then go ahead and get your install discs.

Although self reliance is my watchword when it comes to computers, it does not bother me to call technical support when I hit a gap in my knowledge. Being placed on hold for an indefinite amount of time does annoy me. Occasionally getting dealt the rude customer service representative(CSR) is par for the course as well. Coming across the polite but helpless technical support person is less grating but means a callback is in order. A persistent difficulty in communicating clearly is the defining CSR experience which encourages me to do it myself.

Suppose you are prone to a string of bad habits or a streak of helplessness which makes you reach out to technical support like an alcoholic to cheap beer. Keep in mind that some problems are such that they cannot be fixed by the handy long distance remote desktop feature. Do you really have the time or patience to be on the phone for three hours anyway? Put your hands together to pray for a local specialist with a good bedside manner and great communication skills.

Technobabble is not a real language. Do not let an enthusiast blindside you or a technician confuse you. Where do you start though when "computers are complicated" to you? Start with the fundamental aspect of the whole enchilada. Your operating system needs to work smoothly and properly before you tackle anything else.

If your operating system is badly corrupted you may be sitting there watching the manufacturer's splash screen forever. When it is missing because it was erased by evil spirits, you may see a blank screen with a DOS(Disk Operating System) prompt. Unless you are smarter than the average bear, when you get the plain **C:\>** or some confusing looking message screen about the hard disk, now is the time you call for help. At this point your options are pretty slim so hopefully you had everything backed up recently. Should you be working on a computer older than a kindergartener and backing up your valuable data is not part of your regular routine, you are tempting fate. The chances that you will be able to repair one of those bad crashes are about the same as your surviving a myocardial infarction. A heart attack means fire bad.

Take a few steps back in time to when you got a computer fresh from the factory. The OS was free of the corrupting effects of your flagrant lifestyle of fast cars and loose women. Sometime while you were web surfing, emailing, installing programs, and ~~taking candy~~ copying files from strangers, there have been little changes to your operating system. No doubt a lot of it has been under your very nose, as you blindly clicked "Next" on those updates just so you could proceed to the end without reading what it said. Another time your curiosity made you click through the advertisement/fake warning in a brief moment of weakness when you should have just closed the window. On the street when you see a beggar, do you yield some money, politely give a cordial refusal, or walk on by as you avert your eyes?

Then again, your operating system was never perfect to begin with, so the pristine boy in the bubble routine is not exactly feasible. It was released with problems because like all massive catch all projects, it was designed by committee. Some issues were not even foreseeable until months after rampant testing by a number of different users in a plethora of different environments. Supposedly. The next time you get a golden ticket for a visit to the big software factory, you can ask for yourself. Those updates are there to patch holes all along the bow of that humongous unsinkable ship which floats really well most of the time. Manufacturers love the automatic update feature since it serves to protect the uneducated user and suck up resources at really inopportune times. If you are big enough to tie your shoes by yourself and remember to brush your teeth, you can handle the responsibility of checking for updates at your convenience.

Although modern personal computers are designed to multitask rings around its predecessors, it is always a good idea to just do one type of update at a time. Practical experience will tell you that the same person cannot do brain surgery and play the piano at the same time. Sure it is theoretically possible but the odds are high that you are going to get blood on the keys or screw up a chord. On the rare occasion you end up installing a bad update that got past quality control, you have a better chance of doing a successful rollback if you are not juggling multiple programs.

Often when your computer freezes on you, that horrible state can be traced to something you did or something you neglected to maintain. Some people are still surprised when they get sick after years of junk food and no exercise of any kind. We can easily come up with some parallels between your body's immune system and your computer's OS. Although there are some genetic diseases which are a mystery to us, the workings of a personal computer are not. Think about the last few things you did prior to seeing the blue screen of death, the neverending green ribbon, the kernel panic, the black multilingual you-have-no-choice-and-must-restart page, or the perpetually frozen screen. There was a single piece of straw that pushed your overtaxed system to this moment of cold bleak failure.

If you are able to restart the computer to get to where your computer can run with some stability for troubleshooting, your luck has held out. Something strange happened, some invisible conflict occurred with the last program you ran. Out of morbid curiosity, my instincts goad me into poking it with a stick again. Will the system crash by doing the exact same thing as last time? It could be easy to find the culprit but you know it could also make things worse. Thankfully, everything was backed up recently and there was not too much new since then anyway. This is the perfect opportunity to learn how to prevent this the next time around assuming the experience does not lead you to renounce your electronic servants.

In computing hell, there is usually corruption, conflict, or a problem with a missing file(or several). Any kind of program can have parts of its code corrupted, or rewritten against its original function, much like a tumor; if it is benign it will slowly make the software useless and if it is cancerous it will spread. Conflicts are an annoying puzzle because the inner workings are beneath the surface and you only get to see the outcome of the fight; you have got more programs, processes, and operations going on at the same time than your machine can handle. Replacing a missing file sounds simple except you have to know where to find a clean one and where to put it after you get it; luckily for us, software makers and archivists have pretty good repositories on the internet where you can get what you need.

Although all files are created equal in the eyes of a programmer, the computer sees some as more valuable than others. To hinder your digging around or at the very least protect them from liability, many warnings are in place around the operating system files. Respect your system files and treat them with the same care you would with fine china. Hopefully you can overcome any anxiety over them and actually go see which directory these sets of files are located. When enough system files go bad, that is when you are looking at several hours to replace everything from scratch; whereas non-system software can be easily uninstalled or deleted manually and reinstalled with nary a care in the world.

An attack on your computer can be on any active program you own but the prime target is your system files. Some hackers will create viruses for the sole purpose of publicizing their nickname through the use of a splash screen infiltrating your system's startup files. Others want the bragging rights for taking down the most machines around the world in as little time as possible. Accessing your personal computer and converting it into a zombie slave is actually more fun and profitable. By turning your computer into a harvesting drone, it will collect your financial data and trick thousands of unsuspecting computer users to yield their private information as well. These pieces of software will lurk in your system without any visible interface and run quietly in the background so you may never know that you are an accomplice to a crime.

An up to date anti-virus program is important for not only keeping your computer safe from a massive crash but also to safeguard it against becoming a pawn of an evil computer mastermind. This may sound like a melodramatic cartoonish threat but it is a real concern for the business community. We can personally identify with the lost time associated in having to restore our credit rating but the financial impact to both you and your banks is an exponential nightmare. Given that there are several free anti-virus programs available on the internet there is not a valid excuse for having your computer unprotected.

Knowing the way things typically go for you on the computer, you may even have worms. As defined by your high school biology teacher, a parasitic nematode is a kind of prolific microscopic worm that replicates quickly, causing death and disease in its host. That is pretty much what it does on your computer too. Once you contract it on your hard drive, it can spread to your USB flash drives, removable flash cards, network drives, and any computer connected to it. They can be detected and removed without any change to your system, assuming the infection has not too severe. For anyone familiar with termite infestation, you must know that if it is really bad, the house is going to have to be razed to the ground; a clean slate on your computer is equivalent to formatting the hard drive.

No single anti-virus program is perfect so you should never rely on it completely. Using multiple ones from different companies will just slow down your system, be annoying to manage, and possibly cause software conflicts. The one you choose will be vulnerable prior to the moment you get a virus definition update for the newest baddie. Unfortunately the nature of the human mind tells you that there will be a constant stream of fresh troubles out in the wild.

Active protection against an incoming hostile takeover is exactly what a firewall is designed to do. It is the border guard monitoring the programs accessing the outside world through your wireless device, your ethernet connection, and even your trusty modem. Your operating system probably has an included firewall and you can also install separate software solely for that purpose. A regular router has a built in firewall hard wired as part of the circuitry, which is more robust than using a software solution. Managing the settings for file sharing, remote desktop, or any other kind of advanced networking requires you to do your reading homework. Bear in mind that although you allow your email client, web browser, or P2P to use the internet, it is often through those programs that malicious software gets into your computer. Your loyal bodyguard will take a bullet for you but a pretty assassin will be invited to your side.

Internet Explorer, Firefox, Chrome, Safari, Opera, or whatever browser you are using today serves all kinds of web content including the nefarious variety. They block a certain degree of adware, spyware, and other pop up annoyances depending on what version you are using. However they can be circumvented by adware posing as something useful that you need to install onto your system; these persistent sales gadgets will goad you into checking out their shopping services, weather reminders, adult pictures, tool bars, and false promises to clean your computer. Later releases of your favorite browser will be better able to protect you against yourself.

To protect both the kids and unwary adults against web surfing dangers, take a look at http://www.netsmartz.org and http://onguardonline.gov; although these sites originate in the United States, they provide useful tips without losing much in a cultural translation. They provide you with resources for the personal perils of engaging the gritty online world rather than the technical side of it. Admittedly it is easy to forget there are persons on the other end of that internet connection when you are in a closed room. Just as there are those waiting to hurt, rob, or manipulate you, many others are eager to help, heal, or rescue you. Cue Eurythmics.

Your email carries an enormous set of potential risks that can cause you a long lasting set of headaches. The standalone email client you use could crash, causing you to lose every single message you have ever sent and received; if your hard drive fails on you that is the same thing. Someone might gain access to your email account and snoop on your financial records, personal letters, and complete contact list. On any given day, you could be subject to hundreds of unwanted junk emails, fake charities, offensive ads, and fraudulent ones posing as your bank. Managing the thousands of legitimate work and personal emails between your various accounts requires an orderly system otherwise you will drown.

Preserving your email integrity is going to take diligence on your part so get out your highlighter for the bits relevant to you. Back up your email on a regular basis if you have a standalone email client like Outlook or Eudora installed on your computer; you can choose many tutorials online from concise bare bones directions to step by step screenshots to friendly narrated videos. As bothersome a habit as it is, there is a definite benefit to periodically changing your password to decrease the odds that someone will find a way into your email. Use a dedicated junk email account, a spam filter, or a white list(that is a setting where you can only get email from a list of people you specify) if you really get inundated by floods of email. Depending on how cluttered you keep your mail folders, how full your inbox is, and how many emails you get a daily basis influences if you should organize your email once a month or more often; if you are really pressed for time and your email administrator is threatening to cut you off, sort your emails by size and start culling from the largest ones first(it is going to be the ones with attachments which will push your mailbox limits).

Suppose you have fallen off the proverbial cliff and need rescuing from your email disaster. Losing your password is a matter of contacting your email provider to get them to reset it for you as long as you can prove you are who you say you are. Losing your identity through internet fraud is one of those times where the FBI(or Interpol for outside of the United States) can be your friend; the Federal Bureau of Investigation has an online help center at http://www.ic3.gov. For the lessor aggravation of death by a thousand emails, the FTC provides some information at http://www.ftc.gov/spam for the Americans as well; Europeans can check out http://www.euro.cauce.org . Regardless of which continent you reside, you can check out the Australian government's educational web site about email scams located at http://www.scamwatch.gov.au/.

Keep in mind your computer is also as safe as your relationships with your family, close friends, and neighbors. Earlier the threat of strangers was discussed but you have to think about the trouble your familiars can cause. Should you be going through some trust issues with your wife, husband, boyfriend, brother, girlfriend, sister, father, or mother, one of them could be using keylogger software to track everything you type into your personal computer. Maybe she is just quietly checking your web browsing history to see how many porn sites you have been looking at in your spare time. Your kids could be downloading all kinds of curious things without your knowledge while you are at work, asleep, or out shopping. Of course, the least savvy member of the household is going to be the one to click on the adware/spyware/virus by accident, charge up the credit card after you set it up the quick payment mode, download a few hundred songs using one of those legally ambiguous P2P things, install a few dozen games, and change the desktop settings around without knowing how to change them back.

Then again, you could be the one who is the weakest link. Until you bring your skills up to speed, the best thing you have to rely on is some good sense. With the sheer potential of today's computers, throwing money at it when something is wrong will land you in the poorhouse; should your wealth afford it, you can add technologist on staff to join your personal chef, trainer, maid, accountant, and attorney. Spend your resources on enriching yourself, so that you are able to deal with any machine that comes along.

* * *

For everyone who has made it to the end of the chapter and to those cheaters who skipped to the end, this is your bonus. This is a triage list for the number one complaint from everyone who has every stared at a computer screen: *Why is my computer slow?*

Show this page to someone who can explain in detail anything unclear to you.

The Triage List

- Check the age of the computer[less than 2 years old should be as fast as new, up to four years old requires middle age maintenance, and older than that means you are lucky it is still alive].
- Check the amount of RAM you have[on Windows XP or Win 7 there is a big difference between 512Mb and 1Gb, on Windows Vista 2Gb is the starting gate, with a Mac the more the better].
- Check and clean your personal computer regularly for viruses.[let it run overnight on an older computer so you do not have to babysit it].
- Remove programs you have never used and probably will never need[this is where you bribe your computer friend with an elegant dinner for translations].
- Defrag your hard drive[there is a free program to do this is included with your operating system so feel free to do this every few months].
- Turn off the startup programs you do not really need[see dinner comment above].
- Clean up your registry[dessert too for your friend but Mac & Linux are exempt].
- Get a faster internet service[if streaming video still stutters, then your computer is showing its age].
- Google for additional tips[use the www.google.com search engine to obsessively gather more minor tweaks but this is very very very optional].

Buy Moore Stuff Today
9

Once upon a time there was a boy a named Jules who dreamed of whimsical machines that could do all sorts of magical things. Tiny microscopic tanks swimming through your blood vessels and shooting at cancers deep inside the brain. Supersonic planes, bullet trains, and embryonic automobiles carrying us faster than any cheetah could sprint. Instead of cave drawing to tell stories, the walls themselves would display moving pictures made of light and sound. Shiny portable devices would allow whole orchestras to fit in our pocket, libraries to be compressed into a thin book, and entire museums toured at our own desk.

Although 19th century author extraordinaire Jules Verne is the ultimate visionary for today's modern times, the computer age has been brought to you by Gordon Moore. Gordon is arguably the most famous member of the "Traitorous Eight," a group of engineers who founded several of the core companies of California's Silicon Valley; the origins are better suited to the PBS special *Transistorized* than a Francis Ford Coppola flick. As a cofounder of the Intel Corporation, he epitomizes the driving force behind technology's thinking mechanism, the CPU, a computer's brain. He is immortalized by his peers and history itself for his observation about the progress of integrated circuit development.

Moore's law is based on informal statements Gordon made in a 1965 edition of Electronics magazine. He surmised rather accurately that the complexity of integrated circuit manufacturing would roughly double every two years. Coincidentally, this two year cycle has also been applied to the decrease in cost for computer memory, the increase in RAM capacity, and the decrease in cost for computer data storage. As of 2012 the rule is holding steady and the physical limits of the universe have not slammed on the brakes.

What this means to you is that computers are getting faster, cheaper, and more powerful at a predictable rate. A distinct generation shows up every two years, like superchildren who outshine their aging parents. Although a computer's lifespan is measured by its inevitable breakdown, it can live as long as it is useful. If you are one of those early adopter junkies who simply wants the newest, shiniest, and fastest one on the block then be aware your treadmill is on high.

When computers first started out as monstrous mainframes occupying large rooms, they were literally dinosaurs. Remember those creatures with relatively tiny brains and gigantic powerful bodies, who roamed the Earth over 200 million years ago? In terms of technology, those early computers represented the dominant electronic lifeform less than a century ago. Instead of consuming mass quantities of food, these hefty machines crunched enormous amounts of numerical data compared to an average person. Several theories abound for the extinction of the big boned wonders but we know that a few of the mainframe's successors are still around.

For much of mankind's recorded history, we have been doing things by hand. Those wonderful opposable thumbs have been a godsend in our use of tools. Whether it is due to an innate laziness, a natural creative streak, or a combination of the two, we have built a wealth of inventions to make our lives easier. From the progress of the past we have been able to continually build upon the knowledge of previous inventors, designers, and thinkers. Skipping to the early part of the 1900's via our textbook time portal, the infamous industrial revolution paints a prologue to the story of computers. Machines at this time were pretty much composed of moving parts and people powered like on Gilligan's Island. It was the harnessing of electricity as a power source which brought us out of the dark. This Promethean white fire was also very unique in how it behaved with combinations of different materials. A good physicist who could build things was more valuable than an animal trainer at the circus.

The 1940's was the decade in which researchers were implementing the idea of a programmable machine rather than a one trick pony. At this evolutionary point the transition from calculator to computer looks like the summer home of Boris Karloff's Frankenstein. Elaborate series of mechanical relays and switches would be replaced by vacuum tubes(imagine a glass bulb in the shape of an oversize tampon). It was the monumental leap from a spring loaded mousetrap to a digital watch.

In the next decade we would see the widespread use of the transistor(one might consider it like a brain cell) as a dynamic replacement for the vacuum tube. A transistor was about the size of a pea, physically tougher, easier to manufacture, and it used less power. The prolonged lifespan of a transistor was beneficial too for those who demand constant 24 hour usage. By this revolutionary step, the reliability of a computer became a marketable feature.

A few large companies, the US Federal Government, and some universities were the primary customers of these super calculators in the 1950's. They were really the only entities who could afford to buy, maintain, and use these machines. These computers were exclusively the domain of scientists and programmers who used paper punch cards to talk to them. At the time, the only game you could play of solitaire would have to be with a real deck.

Financial institutions, commercial airlines, and the legendary NASA(National Aeronautics and Space Administration) were utilizing computers on a wide scale by the 1960's. Integrated Circuits(ICs) had entered the industry and cast traditional transistors to the blur of antiquity; ICs were literally transistors finely printed on a thin sheet of glass, enabling computers to further shrink in size. Disk drives were emerging as a way to store and access data quickly and at ever increasing capacities. The first incarnation of the UNIX operating system was also born; it probably beats Latin for the number of offspring it possesses.

Fast forward to the 1970's and the latest electronics magazines showed you how to build your own personal computer like a techie's hot rod shop. These little computers would take up your entire desktop and still set you back a few hundred dollars. It would be really helpful to have a background with basic wiring, a detailed knowledge of electrical components, and an intermediate understanding of schematic diagrams; an architect has blueprints to build a house and an engineer has schematic drawings to dissect a gadget.

From the baby nursery of home technology,

Commodore, Atari, Radio Shack, and Apple were selling their personal computers to ordinary people! Email made its consumer debut for the very first time as does the primeval video game called Pong. Over in England, the addiction more popularly known today as multiplayer online gaming was also created. To win a random bar bet, a Trivial Pursuit pie, or Double Jeopardy Daily Double you should know the ethernet connection(remember, it looks like a phone jack but wider) used in networking was invented by Robert Metcalfe.

In the year 1980, the first hard drive for the personal computer came out with a whopping 5MB capacity; less than thirty years later, you can easily purchase one with 500GB storage which is 500,000 times larger. That same year Ronald Reagan was elected for his first of two terms in office as President of the United States of America. The Summer Olympics were held in Moscow, Russia in what used to be part of the USSR(Union of Soviet Socialist Republics) or the Soviet Union. Martin Scorsese's immortal film Raging Bull starring Robert De Niro is released to the world. Living legend Alicia Keys is born.

During the rest of the decade, the popularity of the computer skyrocketed to infiltrate homes, schools, churches, and businesses across every demographic. While the kids were feeding quarters into a legion of arcade games, their parents were creating spreadsheets at the office. In the movie theater, Tron debuted as a huge milestone in Computer Generated Imagery or CGI. Compact discs were replacing vinyl records and cassette tapes as the way songs were purchased, foreshadowing the mass migration of music to the personal computer. Sharing things between classmates, colleagues, and friends was subtly introduced into the computer culture by the easy to use floppy disc, which was easily about the size of a few slices of baloney.

The 1990's brought lower priced yet faster desktop computers, functional portable notebook sized computers, and the dawn of the internet age. [Read the last few words again slowly and say them out loud, "...the internet age."] People started communicating with each other from across the world regardless of race, income, religion, or political party through a computer. Rules for personal etiquette, legal policy, safe growth, and applicable jurisdiction are being written along the way. Just like the wild wild west, there were newly minted millionaires, outlaws breaking social barriers, and fresh opportunities to alter the way we live. All this arose from a machine which connects six billion people together to share their ideas, hopes, fears, and collective dreams.

In the first decade of the twenty first century, Skynet became self aware. Within milliseconds, the artificial intelligence computer network launched a nuclear attack against Russia triggering a simultaneous retaliation. In the next few minutes, over three billion human lives perished by atomic fire. The survivors immediately faced another nightmare, the war against the machines.

Just kidding.

Assuming that apocalyptic future remains a work of fiction, the new millennium promises to be as bright and intriguing as a gold penny. While nations dominate the news with their military atrocities, the salvation of the individual depends on how well he can prosper in his personal environment. The practicality of the humble computer proves it to be the premiere tool for enriching his lot in life, expanding his reach as well as his grasp.

On a planet where poverty is still an everyday occurrence, not being able to afford a computer puts you at an added disadvantage. An empty stomach is a more pressing concern than deciding what kind of wallpaper to use on your imaginary laptop computer. The same outstretched hand which brings sacks of grain, donated cans of food, blocks of cheese, WIC supplements, and hot meals should also bring digital deliverance. Opening the internet gateway to everyone with a pulse is not some wistful altruistic fantasy. This daring mission of connectivity can be achieved with normal technology at yesterday's soda fountain prices.

Economies of scale have already brought computer costs to a historical low point while maximizing the technical features available for the average person. Spending more money only equates the higher capacity or speed demanded by specialized users. Movie makers, hard core gamers, research scientists, professional engineers, medical imaging technicians, and server farmers are the minority of people who yearn for more than the bare minimum. The cheapest computer on the market is shockingly powerful enough to educate and entertain you for a lifetime without breaking a sweat.

A personal computer has become so commoditized that the brand you buy is less a matter of selecting the best quality and more a statement of preference. Your desktop computer's core components are more interchangeable between manufacturers than the car you happen to be driving; as long as the generation of the computer part matches, the brand is irrelevant. Choosing the color of a new laptop is a curious luxury, which, like the color of the car we ride in, emphasizes feelings over functionality. Despite that a computer is primarily a carrier of content, there has been a surreptitious rise in the styling of a notebook's exterior design scheme; many of the portable computers do not look as ugly and blocky as earlier models. Crossing over from a heavy beige utilitarian box to a sleek work of art marks a significant transition from practical to pretty.

After you finish your Candyland daydream where everyone is rich enough to have a glossy superthin laptop, the realistic way to connect everyone is to share a regular desktop computer. Parsing the time between the number of users per computer is an age old resource issue which can be resolved with proper time management and/or more computers. From a technical standpoint, having three or four user profiles on a family computer is no big deal; the operating system creates user settings in each person's own customized directory. Although a typical personal computer theoretically would have no problem with a classroom of thirty profiles or three hundred, the hard drive space would need to be managed carefully; should the drive fail, you are looking at the collective loss of everyone in the group. Configuring a server to manage a network of users(set up a master computer to connect as many client computers as you need) would be the next logical step but the sheer cost and technical training to maintain the system is more appropriate for a growing business. The best way to share X number of computers with Y number of users across a Z number of system setups is to give everyone a portable USB flash drive to keep both their personal programs, documents, and misc data.

Oh be it ever so humble, there is nothing like a USB

flash drive for freeing a person from the bounds of one's petulant computer. Suddenly your computing luggage has dropped from six or thirty pounds down to the feathery light form of a disposable cigarette lighter. As long as you can use someone else's computer, you can carry around a digital soul in your pocket, on a necklace, in a purse, or concealed in a taboo bodily orifice. The cost per unit of memory drops too quickly for us to publish numbers with any sense of meaningfulness. Let us just say we can already carry on one drive a small library, a singer's discography, a full fledged operating system, the tools to diagnose and cure an ailing personal computer, several movies, or every spreadsheet you ever have compiled in your meager career. Whether you are merely a student or teacher, consultant or client, traveler, tinkerer, or spy, this is your new swiss army knife.

The altruistic push for "more computers" is way mightier when directed toward increased computer access for people rather than the mindless ambition to buy more machines. As fast as these personal computers are breeding now, an accelerated roll out to the digital deserts may be unwise. Right now the vast majority of people who use a computer barely master them as it is. Some quality time would be great rather than just spending the day to catch up on television reruns online. Wasting away in front of the screen is an idle luxury afforded to much of the urban internet crowd.

As a spoiled technocrat of the electric caste, the new religion of instant gratification fuels our appetite for shiny pretty plastic things. One tiny scratch on the side of the case and we are ready to look for a fresh replacement. A colorful stylish silhouetted billboard makes us drool for the fun happy toy, price be damned. Those friendly people on the infomercial are saying this is the best computer deal they have ever had on the show and the *limited quantity* flashing on the bottom of the screen is causing the phone to beckon. Online shopping is a shopaholic's perpetual motion addiction brought to a twenty four hour anytime anywhere nicotine rush. All we need is a plastic card with sixteen numbers brandishing Luhn's algorithm and a blue sky credit limit. After a few mouse clicks, a couple confirmation emails, some impatient days, and a scribble on the delivery guy's worn out pad, your virgin plaything arrives in an oversize cardboard sarcophagus.

So how much corrugate have you consumed in the past year? Empty boxes sit in your closets, up in your attic, down in the basement, out in the garage, and quite possibly piled up in your living room. You have kept them just in case you need to return it to the store, mail it back, resell it to someone else, or to warehouse other stuff you bought last year. By stockpiling those boxes you have provided a nice home for an array of spiders and insects including moths, beetles, earwigs, centipedes, cockroaches, scorpions, and ants. As the rare species of diligent recycler or neat freak who immediately disposes of his/her pulp packaging, the evidence has just slipped out of your line of sight.

Then again there are the half broken pieces of electronics you have managed to abuse over the years. Think of all the battery powered gadgets and gizmos you got as a gift, won as a prize, bought for yourself, borrowed indefinitely, or just plain stole. Add in the laptops, car stereos, sound systems, televisions, VCRs, desktop computers, and various components you upgraded as well. How much do you have cluttering up your home right now in some haphazard state of repair, fast tracked for the trash? When your kids decide to create a three dimensional tornado collage of all the cords, plugs, cables, adapters, and remotes they found, that is a burning sign of consumer excess. A simple historical snapshot of the electrical machinery you have fondled in the last five years paints a vivid picture of modern conspicuous consumption.

As any scavenger knows, this cumulative flotsam adds up, multiplies, and goes exponential as soon as you look around. Garage sales, community computer disposal events, your engineer friend's house, any school's aging computer lab, Goodwill stores, and occasionally a shipping container bound for China will give you the perfect vantage point. Your carbon footprint is a paltry thing compared to your heavy metals shadow or your plastics disintegration factor(let's call that the amount of plastic you use over a lifetime multiplied by the number of years it takes for it to decompose). Woodsy Owl, the "crying Indian," and Captain Planet would probably find themselves overwhelmed at the amount of ecological damage one of us does daily.

Electricity is one of the djinn which Prometheus delivered to us to conquer the universe; at least it is in the version I tell around the campfire. At best we have a basic idea of how electric fire behaves in a circuit, as the strings to our limbs, as a jagged flash streaking down from the sky, as a sharp sting from a static charge, and as a trained snake whose fangs lay recessed in the wall. For the majority of us we see it as a gifted slave who furnishes power for our limitless machines of light and sound. Like a newborn baby it has to be coaxed to where we want it to go or rather carried with insulated kid gloves. Given the hectic mental routine we spin our minds around, our preoccupation with electricity is often limited to whether it is on or off.

Our appetite for electricity is a perpetual hunger that we have to manage closely as we play with more toys. Your monthly electric statement is probably the cheapest bill you have to pay on a regular basis yet it is the most universal. Either mom or dad made sure you had the phantom voice in your head well into adulthood echoing the reminder to shut off the lights when you left the room. Of course no one is going to ration the hours the refrigerator is plugged in or restrict the number of times you can use the microwave. Television's frivolous energy drain is worth having for the mealtime luxury, vacuous companion, background noise, cartoon treats, and dramatic escapism. Laundry machines, air filtration systems, dishwashers, sonic toothbrushes, and other modern conveniences of the post agrarian age are other leeches on your electric grid. That leaves just the village of wattage pirates inhabited by your computer(s), monitor(s), printer(s), router, broadband modem, television(s), DVD player, VCR, DVR/Tivo, video gaming system, stereo(which could be a dorm sized boombox or an elaborate surround sound theater experience with the full contingent of black boxes), and chargers for your mobile phone, bluetooth earpiece, pda, rechargeable batteries, mp3 player, portable DVD player, digital camera, video camera, electronic book reader, antique pager, and your kid's handheld game console.

Unlike Imelda Marcos' shoe collection, your fondness for collecting electronic gadgetry has not gotten massively out of control quite yet. Nearly every electrical outlet does happen to have a power strip snaking out of it, strung up with ebony strands of plastic licorice. Those blocky black adapters cover large areas of the power possum like some unnatural fungus wanting to be plucked. We will be extra careful not to trip on any of the cords, spill anything near it, or move things around without telling you.

Wishing for a wireless world is a nice sentiment but empowering up the armies of laptops is going to take an inventive breakthrough. Portable devices are either plugged in every night for bed or they poop scads of alkaline batteries with the sheer regularity of a fluffy bunny. Double As rule the popularity roster as shapes and sizes go, although the variations you can purchase mimic the diversity of rocks in a desert; finding the exotic battery you need is aggravating enough to drive you to pay just about any price the store demands. Rechargeable battery technology works well enough for a few years but the decay of entropy is annoying enough to piss off a casual user; lithium and nickel based electrolytes(the jelly in the donut) may beat the pants off disposable batteries but even they need to be replaced. Promising discoveries in materials science engineering have not yet made it from the laboratory to the factory floor despite the nanoparticle research you found on the internet.

Once we resign the electrical umbilical anchor to the death and taxes category, we can look at the other intangibles we are paying for to make Mr. Computer happy. Do you get the two hundred dollar warranty(which does not cover your physical computer *or your data* against theft, accidents, fire, earthquakes, or other acts of God) on a small machine that loses half its value in a year? Are you going to get extra online storage for your personal files so you can access your archives while traveling? Do you have a nice budget set aside to pay for your web site's domain name, alternate domains, setup fees, hosting fees, design costs, maintenance costs, and search engine placement? Have you already spent the money on anti-virus software(plus the yearly subscription for updates), spam filters, spyware blockers, and a firewall to monitor data traffic? Will your sense of paranoia force you to pick up some heavy duty backup software to create disc images for archives, boot discs in case your OS crashes, and separate partitions to maintain your own private virus zoo? Perhaps the default operating system that comes with your new computer is so incredibly annoying, that you would gladly pay for a different flavor.

Those expenditures are just to get your vehicle insured and registered with the DMV, so to speak. To do anything truly productive, you are going to need an office suite composed of a word processor, spreadsheet software, a presentation program, an email client, and perhaps a database program. For the down and dirty desktop publishing arena, you might want to drop some cash on a program or two that can generate flyers, brochures, business cards, invitations, and greeting cards. There is an enormous variety of vector graphics and painting programs floating out in space for creating professional level artwork or crayon like doodles. Image editing software would be great for cleaning up your digital photos with glaring wrinkles, red eye, blemishes, love handles, or those obnoxious passers by. A decent sound editor would be useful to roll your own ringtone, remix some songs, record a birthday greeting, send a vocal love note, or to make some special effects for your answering machine. If you are one of those home movie people who must film the first few years of your kids' lives in video so you can show them in high school how cute they once were, you can add a video editor to your shopping cart. Do not forget the money management software to help you manage your finances and a separate one to help you file your income tax. [Note there is no mention of video games or entertainment because you can seek them out all by yourself, price be damned....]

Admittedly you may not need every single one of them but there is a common question for each of those programs you are going to have to figure out before you fork over a credit card. Which one are you going to get? Obviously, the stock answer of "I want the best one for the lowest price" is going to earn you a polite stare and a gentle pat on the head. Any type of software is going to have several products competing for your hard earned buck and what it should boil down to is buying for your needs. All of the them are going to have the same core features but the defining factors are the user interface, the extra features you specifically want, and how stable the program is. Remember it is your prerogative to get what you need while refusing to settle for only what that one particular store has to offer. Researching online for a piece of software categorically is the homework that makes you a great shopper.

Practically every software on the market has a trial version, a description of the minimum system requirements(like the "you must be this tall to ride this ride sign" in front of the line at the roller coaster), a list of key features, screenshots of the program in use, and user guides available on the manufacturer's web site. Usually the trial version is full featured for a month or less and/or it has certain key features turned off until you type in the special key code that says you paid for it. Those system requirements are especially helpful when you have a computer which is older or falls outside of the most common configuration; this is never something you can ignore but your next software installation will be smoother after you discover you have a typical computer. Between the marketing spiel and the engineering specifications listed on the features list, you had better find yourself a great translator and/or have done the homework I mentioned above. The screenshots are sometimes helpful in showing you if the programmers made the software intuitive to use without needing to go for a test drive; you do walk around the showroom at the dealership and sit in the cars, right? Depending on the diligence of the manufacturer, you will find a range of help material ranging from a digital copy of the manual to detailed video on how to optimally use their product.

Either you have gleefully dropped down those thousands of dollars for all that software or are bitterly contemplating how much the rest of what you need will cost. It is a no-no but bootleg copies can be found at severe discounts although the viral risk to your computer is worse than getting caught. My gut tells me letting a friend borrow your installation disc occurs more often than loaning someone your car but that is practically impossible to prove. In many cases there is a watered down version of the software you want already bundled with the computer. It is not necessarily packed chock full of features but it is functional; any new device you get like a printer or digital camera will come with diluted bonus software too. Another avenue is to find a slightly older version of the software you want, which should work fine and be cheaper to boot; last year's car model is probably going to get you from point A to point B just as well as the newest one right off the lot. Then there is *open source* software...

Open source software is distinct because it is both free of charge and its source code(the precious set of instructions written by the programmer) can be openly viewed, modified, and republished by anyone. Once upon a time this collaborative concept would have been just a brilliant hippie's pipe dream. Fortunately for us, the idea has spread to reflect practically every piece of commercial software available(deja vu). Ambitious enthusiasts are providing enough momentum so that newer versions are constantly being produced. The maturity of a lot of open source software is so high, it rivals the profit driven version. It not only provides an immediate cost advantage but in many instances the community nature of its development yields a more stable, secure, universally compatible, and user friendly program; progressive companies are expanding their usage of open source software both for their server network infrastructure and the applications used by their employees.

Obtaining open source software from the internet is the primary way to get the latest version of whatever you need. Although a store could sell a CD/DVD with one or more programs on it, the speed of new releases would quickly render it outdated. As the nature of open source software tends to make it much leaner, downloading it is not that big of a deal even on a slower connection. Detailed version histories online show the value placed on a software's documentation rather than its marketing. If right now the open source version(s) does not have the exact features you need, check back in a few weeks or even a few months.

On the printed page right here, I am going to blithely rage against the infamous "all computer books are born with zero half life" rule and provide a few open source software titles you can find and download for your personal benefit. Mozilla Firefox (tastes like Internet Explorer) is a swell web browser enabling you to surf the internet. Open Office (smells like Microsoft Office) is a full featured suite with a word processor, spreadsheet, presentation, and mathematics program rolled into one. Gimp (almost like Adobe Photoshop) is an image editing program. Mozilla Thunderbird(a standalone email client) is for sending, receiving, and managing your email. Inkscape (similar to Adobe Illustrator or Corel Draw) lets you create all kinds of drawings. Kompozer (rivals Front Page) is a web page editor which lets you build your own web site. Irfanview(proprietary freeware and not open source but it is written by someone out of the kindness of his heart) is an universal image viewer.

The Gnu General Public License(GPL) is at odds with commercial software companies, waging battles in court underneath the public radar. A few profiteering CEOs view the socialistic sharing of software as an abomination, a threat to competition. As a knowledgeable user, you can bask in the benefits of the Gnu regardless of what the big bad wolves says. [In the future, there will also be free web based applications to further displace the expensive store bought software. You can see it sprouting up but it will not blossom until broadband internet reaches everyone in the global village.]

To buy Moore's stuff we don't always need to buy more stuff.

You Can't Escape From RFID
10

You cannot escape from RFID or Alcatraz. The latter is just a former high security prison located on a tiny island in the San Francisco bay with an infamous reputation for eternal incarceration. The other is a simple radio frequency identification tag. This tag is a tiny flea sized microchip which can be clipped to a product in the store, sewn into a backpack, glued to a corporate entry badge, hidden in a broach, or injected under the skin. Acting as a wisp of a radio transmitter, it passes on information every time it gets close to a sensor. Your location can be tracked on a computer monitored floor plan, building schematic, or topographical map like a rat in a maze.

From a social perspective, the advent of RFID establishes the day electronics crossed over from beeping toys to tools of the nouveau damned. While Elliot Richards, Buffy, Ash, John Constantine, and a fiddler from Georgia fought tooth and nail against the forces of evil, those electric gadgets are slowly bleaching our souls. The moralistic sins of yesterday may even be losing ground to the antisocial plague of stoniness; more and more care less and less. As drones of flesh we deftly plug ourselves into a computer terminal or portable device and disconnect from the world. Thanks to the invisible leash of a RFID, our transformation into live statistical data is complete.

The products we consume are being fed into redundant databases every second of every day. Your habits are being tracked to create a demographic profile for the sole purpose of selling you more stuff. Companies salivate at the prospect of generating a profit through the use of predictable patterns; in fact you are being hunted down like an endangered migratory bird trying to protect its nest egg. By reaping the smallest details of your shopping preferences, marketers can send you a faux invitation advertisement of your adolescent dream car playing your spouse's favorite songs on your anniversary. Someone like you wants to buy exactly what we sell each and every day, week, month, and year for the rest of your anticipated life expectancy.

For good or evil, practical or superfluous, the devices we touch have the potential to deeply benefit us or to rot our brain, empty our wallets, rob us of time, cause a great deal of stress, and clutter up our living space. Conceding to those negatives and the big brother factor, it is a noble challenge to seek the good in an electronic gadget. We want the thing to be useful for more than merely bragging rights; your wide eyed plea for the smallest and sleekest, the latest and greatest sets you back two steps.

What digital toys have you got plugged in to your psyche? For the sake of crazed optimism, we are going to skip both the bad and the ugly. This time the arrow is aimed straight for the positive...

On a good day, a digital camera can enrich a fun experience, preserving a moment in time for easy sharing with others. Your average point and shoot model can easily store a thousand pictures as sharp as any postcard at the local tourist shop. Intrinsically the images it produces are more pliable than a chemical film stock, allowing you to infinitely bend and twist them on any fresh computer. With just a little help from the internet, it makes every amateur photographer an international journalist, an accidental artist, the family historian, a potential paparazzo, and a corporate spy. For using such a remarkably ephemeral media, it helps to harden our own fragile memories quite well; their digital images reproduce snippets of life to aid the mind affected by Alzheimer's, autism, a stroke, or the plain undiagnosed bout of forgetfulness.

Our choices in the marketplace for form factor are now a matter of preference since the majority conveniently fit in a small shirt or pants pocket. Anyone old enough to remember looking through a viewfinder(the little glass hole on the back of the camera you used to press your eyeball against) to take a picture can rant at how easy kids have it today with the fancy LCD screen. Being able to review a shot instantly after it is taken and erase a bad picture is the crucial selling point for every style conscious princess. Another blessing from the electronic camera gods is the exiling of red-eye and blurry snapshots to your grandparents' spiral bound photo album past. Today's photographic technology has taken the hobby out of the hardcore shooter's realm and put it into the hands of anyone who

148

can press a button...

Anyone who can escape the gravity well of his hometown to travel the Long Way Round the neighborhood will probably take along a GPS(Global Positioning System) unit. Although they used to be exotically expensive and solely the playthings of the military, an average American soccer mom nowadays easily afford it. Trips outside of the comfort zone are made possible by a digitally recorded voice telling you how to get to the new restaurant downtown without having to pull out any sort of fold out map. Better yet, the global coordinates allow you to imagine where you are on the surface of the Earth, as a slowly moving pinhead speck.

Using a GPS forces you to instantly broaden your understanding of the nature of world geography. When you draw those imaginary lines(of longitude, which are all long) from the North Pole to the South Pole, they always require the same length of string. The fabulous figurative ring around the world's midsection called the equator, is the most universally known line of latitude; you get a gold star for remembering how the various size circles resemble rungs on a ladder when you stare at your side of the world. Those two sets of numbers representing your location bring meaning to the criss cross grid on every globe you ever spun in school as a kid. How it cleverly breaks down into degrees, minutes, and seconds brings back hazy memories of elementary school math and social studies.

Aside from the prolific car navigation device, the technological heart of GPS can be implanted as a secondary feature to other toys. Brought to you by the telecom giants, this generation of GPS fits on a single chip, small enough to be swallowed. Sure, it still needs a power source, some kind of case, and a human interface so we can use the thing. That is still small enough to be hidden inside your fancy boat, private plane, sports car, or customized motorcycle in the event it gets stolen. Buried inside an expensive watch, cell phone, or laptop, it has the added bonus of being within something you already carry around...

You already carry a portable music player too, which may become obsolete in a few years when it gets absorbed by your cell phone; let the idea fester a while the guys in the lab fit your entire music collection on it. Maybe you use an aging Walkman that still plays cassette tapes. Or perhaps you wear one of those ubiquitous mp3 players on your arm while you jog your thirty minutes on the treadmill. Each of those little things clamor for some kind of recognition and respect in the shadow of Apple's popular white wunderkind. Even the gorgeous iPod will fade away though, in the bright lights of its predictable successors.

While we patiently wait for our million minute music players, we can relish the nice chunks of sound in our hands now. A typical device can hold several audio books for your listening pleasure, allowing you to pause at any section or skip to your favorite chapter. Whole concerts, studio recordings, lounge sets, and special remixes can all fit snugly in your pocket(hmmmm, deja vu). Those recorded gems such as radio plays, historical speeches, candid interviews, and erotic audio enliven the theater of the mind. Using it to learn a new language becomes part of your daily routine in your secret plan to travel overseas, romance someone, get a job promotion, and start a new life. Your favorite pickup artists, motivational speakers, financial planners, and spiritual leaders have plenty of material to keep you busy for countless hours.

Given your vast experience and expertise in your field, you too can start making homemade podcasts(downloadable audio recordings "broadcast" on the web for any iPod type player) for your faithful clients, rabid fans, devoted friends, and loyal minions. Whether you genuinely have something valuable to share or you merely want to feed your memoirs, you can easily add verbal musing to your blog. This is your opportunity to elevate your cult status from ranting loner to aggravated anti-hero! Or maybe it is enough to do readings of children's stories for the world library; keep the oral tradition of storytelling alive by putting some emotional oomph into it.

Unless your music player has a built in microphone, you can obtain a dedicated device just for voice recordings. Disregard the cassettes tapes of yesteryear and jump right into the solid state [look Ma, no moving parts] recorder. Slip some extra batteries into your pocket to push your monologue sessions past the week. Eventually

150

when the memory card does get full, you can copy over your sound files to your trusty computer in a matter of seconds. Label them for posterity, edit out the long pauses, tag each file with subject keywords, archive them safely, share on your web site, and go back Jack, do it again.

My fondest wish for an audio accessory would have to be an instant conversation translator. Speak a sentence or two into the microphone, dial the language preferences, and seconds later you would hear the translation from the speakers. Behind the scenes, the coders would start with an excellent voice recognition software, combine it with a fuzzy logic linguistics program to adjust for slang, and finish it off with a nice text to speech vocalizer. A light wearable version is not quite yet commercially feasible but the technology does exist for all the pieces to be assembled in a crude clunky form.

For now, we will just have to keep using a living translator or settle for the tedious method of pounding letters; the non-alphabetic languages always pose a challenging keystroke dilemma. Slowly typing simple phrases into a pocket dictionary or internet enabled web translator are our only electronic options so far. Although our fantastic aerial inventions can carry us across time zones with ease, our linguistic instruments can stand a little more product development. At best, the preprogrammed output will be limited to a handful of languages from the same part of the world. A search for mildly obscure dialects will result in blanks stares, hearty guffaws, or outright rejection from even the most resourceful electronics dealer. The phone app is coming very soon to cell phone near you.

In the future, Babel's curse will eventually become either a distant memory or a whimsical curiosity. That parallel universe, where we cultivate a universally spoken language, will make the preservation of a flavorful polyglot a true anarchist's mission. Armed with an enchanted electronic translator, the hero scurries to duplicate the lyrical nuances of Portuguese fado, Creole blues, Navajo songs, urban rap, and Cantonese slang for new world ears. The final place left to relish the richness of the past will be through   scenes on the actor's stage...

The actor's stage has actually expanded to fill every

crevice of the visible world and extend everyone's fame allowance past fifteen minutes. Everyone is a celebrity, a star, a public speaker, a clown, or some overwrought combination. Willing participants include the lonely girls vying for attention, the aspiring singers emboldened by years of karaoke, the daring stunt athletes showcasing their falls, the nutty desktop philosophers rejected by cable access, and the clueless politicians "embracing the internets." Unwary victims of the video age will find their temper tantrums, verbal flubs, unintended nudity, poor driving, and pure idiocy broadcast to the audience of six billion.

For only a few hundred dollars, a lightweight video camera can be mated with your hand to capture everything you witness as proof. As an additional perceptual eye to the world, it provides further access to the other 99.999999999999% of life we are not able to see firsthand. It enables the soft spoken, shy, and inarticulate to overcome their handicap by becoming storytellers. Despite not having the last name of Coppola, Zhang, Spielberg, Fellini, Kurosawa, Hitchcock, or Kubrick, you have the opportunity to create your own cinematic masterpiece.

With your trusty camcorder at your side, you can televise the revolution, somber aftermath, commercial breaks, blistering parodies, and unforgettable public service announcements. By this boon of technology, we diligently observe the outside world as a reflection of our inner self. Every scene we record is potentially a testament to our ignorance, brutality, stupidity, cruelty, and the eternal struggle to choose otherwise. Be sure to film the human monsters carefully and from a distance lest you get caught in the crossfire. Yes, literally.

Variations on the video preservation unit include CCTV(close circuit television), police car mounts, bank ATM cameras, burglary monitoring setups, laptop webcams, dressing room spies, and nanny cams. Fed by an endless well of paranoia, bleary eyed security guards keep a lookout for any number of wrongdoings. Archives are kept on magnetic tape, banks of hard drives, and optical discs for some casual voyeurism. A small half step is all it takes to link all of a company's servers together, so that accessing a single point gives you keys over the entire kingdom...

The entire kingdoms of Narnia, Oz, Camelot, Siam, Midrealm, England, Duloc, China, and France may be encapsulated on a personal video player. Whether it is a strict reenactment or pure historical fiction, several dozen accounts are available at your very fingertips. Folk tales, animated adventures, travel specials, biographical documentaries, political dramas, food tours, and costume design showcases are some of the choices you may select. As a matter of bloody fact, every single video you watch on your computer can be compressed for viewing at your convenience.

Airline flights which sorely lack the entertainment you crave are remedied by bringing on your own shows. Those crowded train rides on your commute to and from work are the perfect place to catch up on last night's late night TV. When you are stuck behind mom and dad on a long car ride, videos are a win-win distraction for everyone. Waiting rooms at the doctor's office and your favorite governmental facility(DMV, post office, immigration services, social security, passport agency) are no problem at when you have your portable distraction. As a member of the impatient nation, any downtime at all throws the ADD into severe overdrive.

Once we plug in a pair of earphones, we belong to the little video player for quite some time. Any sudden loss of power will shut us down as promptly as a cranky baby screaming its lungs out. Until then we can savor small doses of mind candy in a stream of saved sitcoms, queued up for the long line ahead of us. An hour or so of a good drama series may make you a bit edgy during your stay when the goal is to survive the office ordeal. A movie paused abruptly when your turn comes up should be easily resumed even with the most basic type of player.

Those clam shell like players that play DVDs are well loved because they are the easiest to understand and no computer is needed. Most of the others have their own internal drive or accept insertable flash memory; those still need to be attached to a computer to load up. For a hefty price and a steeper learning curve, a cutting edge one will pulls movies right from the ether. The variables of price, ease of use, and size fall into the infamous "pick the two you want most" paradigm.

Your screen size matters a lot when you have a portable

153

media device. Try not to let either the glass-face-could-break-at-anytime or the steal-me-now paranoia disturb you. Once you get used to the matchbook sized picture, the eyestrain will creep up on you before you know it. Scenes emphasizing people in conversation are much easier to follow visually than car chases, fist fights, giant transforming robots, gun battles, or 300 soldiers struggling for their honor. On television, people's faces may look like large dolls but on a pocket screen the nuance of facial expressions are lost to the ant world...

The ant world is humongous on a widescreen high definition flat panel hybrid of glass and electronics. A nature special on the picnic spoiling insect looks like a remake of Jurassic Park. Ultra sharp images appear as crisp as being at the front row of a live play behind a crystalline window. Today's technology casts off the haze of Hollywood's dreamy atmosphere, firmly replacing it with the power of a microscope. HDTV allows the audience to see every imperfection of an actor's immense pores, layers of foundation, puffy skin, veiny hands, and road map of wrinkles.

Home theater's migration from the celebrity level mansion to the middle class suburban habitat is easily traceable. Its subtle transformation from large wooden furniture centerpiece to a framed electronic painting to chameleonesque wallpaper is visibly underway. Cave paintings have come a long way baby.

At the big screen smorgasboard of consumer electronics you will be overwhelmed by the mortgage sized prices, candy shop selection, and various geektastic technologies competing for survival. Although the median MSRP(Manufacturer's Suggested Retail Price) has fallen from the cliffs of insanity, it is still a hefty investment. Elegant black dominates the High Def player's club though you will see the occasional glint of silver; true renegades will customize the casings with the anachronistic flavor of steampunk, a shiny brass oddity from H. G. Well's alternate timeline. Plasma, LCD, LED, Blu-ray, 720p, 1080p, region codes, divx, codecs, laserdisc, VHS, OLED, HD DVD, DVR, HDMI, DRM, dvi, composite video, coaxial, and progressive scan are the twenty bits of jargon you will need to know when you head to the store for your "big screen" test...

Your big screen test is secretly a monument to the modern media library. Daddy's mojo is measured by the number of movies in his personal collection. A colorful row of VHS tapes reflect every Disney film ever released to the general public. Right below the lava lamp and unsolved Rubik's cube, the stash of vinyl records is keeping the laserdiscs from feeling lonely. As we pass by the dusty stacks of DVDs piled on the coffee table, passing thoughts of Blu-ray occasionally creep into the psyche. *Every time the movie companies change formats, it costs a small fortune to changeover my movie collection,* says the ubiquitous bitter consumer.

Transition players that combine two or more video format technologies are a nice blessing for many film fans. From our previous lesson in declining manufacturing costs, we know it is not much of a stretch to marry an old machine to a new one. At the local tech market, we should be able to find the May December couples of DVD/Blu-Ray, VCR/ DVD, record player/USB, 8-track/USB, cassette player/USB, laserdisc/DVD, and 8mm video/Blu-ray. We can simultaneously save the past and embrace the future by preserving analog memories alongside our new digital ones. Any legacy materials found to be fading and degrading, can be translated to the current media. To some degree, your childhood can be restored and repaired close to the way you remember.

One nicety about collating the home movies of yesteryear, is how they can now fit in so much less space. A few crates of VHS tapes are equivalent to a row of DVDs on a shelf or a hard drive the size of a small book. Saving all the old media to your desktop computer gives you back an entire closet worth of space. Scouring a wall of videos has its novelty but if your overflow includes piles of discs in front of the TV, then browsing a list on your computer is much simpler...

A much simpler way to share recorded TV shows is to use a USB flash drive as an updated videotape. Like the clunky cartridges of yore, you can erase and reuse them repeatedly. Your DVR, DVD player, or computer acts as the base station for the flash drive powered space ship to dock into port. Old school video/file sharing is renewed when you see your friends face to face and personally loan them a show you enjoyed. If you really feel like indulging in a marathon session, pop a huge bowl of popcorn and load up a full season of your favorite show on the flash.

Stocking up on more than a handful of USB flash drives is not horribly expensive but it is wasteful for a few other reasons. Not having the time to do a little bit of basic file management is ludicrous even on an slowly functioning computer; when your home computer cannot handle this, you have bigger problems than deciding what to watch. Get the largest capacity you can find on sale with an emphasize on the "sale" part; they will inevitably decrease in price so shop the best you can. Several years from now, these data carriers will become antique trophies as well.

Today the technology exists to control your television/movie experience wirelessly from a run of the mill store bought computer. A savvy user can perform the hardware setup, manage the basics, and show fresh faced visitors how to navigate. We are still a ways off before normal mortals can easily afford the cost, acclimate themselves to the concept, and adopt it for their household. After you invite your tech friendly hobbyist for a night of filet mignon and open source, you can join the early adopters. Once everything is in place, you just need to drag-and-drop before you push play.

Until then, you could invest in a large capacity external hard drive to use as an interim movie library. At less than a hundred dollars for a terabyte, it is a decent investment to store 5000+ movies; this lovely paragraph is time stamped as Spring 2010 for anyone wanting to wax nostalgic. Leave the other one you bought as a backup drive to remain dedicated to that noble purpose. On this one you can decorate it with stickers and happy paint to commemorate this video jukebox; remember to avoid those fridge magnets kiddo. My pantomime to show the number of DVD boxes it replaces would probably be me laying on fifty stacks of ten arranged like a bed; another appropriate image would be me tossing the 500 plastic cases into an oversized wire waste basket using a combination of time lapse video and CGI editing. Add in the progressive scan DVD/USB player for fifty dollars and you will have everything you need...

Everything you need to access your life's electric shadow

will soon be on your mobile phone. It will be a feature rich über device which surfs the internet, voice dials your calls, controls your home theatre system, accesses a credit line to pay for dinner, provides GPS information, records high definition video, and makes delicious french toast. This fantastic cell is almost here but you will be indebted to a large monthly fee. My penny pinching brethren will have to wait about a decade longer before theses toys are found in a cereal box.

Internet enabled tablets and ebook readers are the toys on everyone's must have list. Compared to a laptop, they are lighter, last a lot longer away from power, less powerful, and lack a keyboard. Pretend you ripped off the screen from your laptop and made it touch enabled. Usually you will have a 10" or 7" color slate or a slightly smaller black and white one. Away from a wifi signal or cellular service, you can play your music and games if you can wrest away from your 3 year old daughter.

With one foot in the proverbial grave, PDAs(personal data assistants) are close to becoming an electronic footnote. In the business arena, diligent retail workers still use these handheld wonders to manage inventory on the store shelf. For the hip consumer, it is assimilated into the current generation of smartphones. These palm sized calendar organizers, note sorters, address books, portable solitaire dealers, and email carriers are historically the standard issue of the urban yuppie. Their adoption into mainstream life shows how many more of us are living in compressed time, squeezing as many activities into a 24 hour global day as possible.

Do you remember pagers? They are those little black buzzing things beloved by doctors, drug dealers, stockbrokers, salesmen, and social butterflies. When someone urgently needs to reach you in a collect-call-kind-of-way, he could simply dial your pager's phone number. By punching in your contact number at the recorded prompt without any human intermediary, it has this feel of spyish covertness. With a coy glance down at the friendly vibration on your waist, your self-importance raises a few ticks on the vanity meter. In a fit of techno-nostalgia you may just be able to find a cheap pager along with a working service if you look hard enough.

However, its general popularity long since reached a cultural peak before diving into the pit of obsolescence.

The extinction cycle of every electronic curio you love to fondle is pretty much predictable. On the front face, the tactile surfaces become so worn out that the printing on the buttons fades to mysterious blank pads. Software updates from the manufacturer will cease soon after they kill the advertising. Any device needing a network service from the company to function, faces the inevitable sudden death scenario; when it stops being profitable, the corporate masters abruptly turn it off. Upgrade fever will also throw you into an obsessive *want* mode when you see the glow of a fresh new model. Younger, stronger, and more responsive are what we desire in a constantly craving; you may even feel it seething beneath your skin.

At some point your romance may transform itself from an addiction to a positive relationship of some sort. When you find yourself buying more from necessity than the allure of the marketing campaign, you are in a good place. Give yourself a pat on the back too if you have at least managed to stick to a monthly budget. Your ability to afford these toys is always going to be at odds with your childlike whims.

When we were kids, nothing would match the satisfaction of being able to carry your toys in your pockets or purses. Our brains somehow register a delicate ladybug, a weird looking rock, shiny glass marbles, four leaf clovers, key chains, seashells, and whistles as simple fun. At some point we get our nimble hands on flashlights, radios, remote controls, phones, and keyboards of many kinds. The die is cast. We are hooked on the buttons, the shiny, the colorful, the noisy, the smooth, the light, and the fanciful. From the cradle onward, we grab anything and everything within our reach because it makes something inside of us go wheeeeeeeeeeee.

Adulthood is no escape from our inner child's need for playthings. Our toys may be more sophisticated but they are more expensive as well. Our attraction to them can still be traced back to the same fascination with blowing soap bubbles, building sand castles, sidewalk chalk drawings, playing pretend, and reading under the covers with a flashlight. Our high definition surround sound theatre system is another means for us to engage in some make believe. Sending text messages to people we know is a throwback to passing notes in class. Video games are naturally reminiscent of playing sports but without the actual outdoor activity.

It is okay to fill your home with gadgets galore. It is normal to spend hours at a time, several days a week, and months on end playing video games. Everyone is buying new phones for themselves nowadays; you absolutely need a new one so you have all the latest features. Just remember that the major electronics store sales start on Sunday and new release DVD days are on Tuesdays.

This is socially acceptable behavior. [Just keep telling yourself that.]

A Billion Monkeys Who Like Themselves
11

There is a philosophical thought experiment which suggests that if an infinite number of monkeys on typewriters are given an infinite amount of time, they could produce the complete works of William Shakespeare. With the advent of the internet, we are close to fulfilling a close approximation. At over six billion baboons and counting, we can come up with a lot of ways to bide our time. The majority of the world may not have a computer, an internet connection, or even a regular meal but the conversion is coming along. Clusters of current users have an amazing amount of power creatively, politically, and financially whether their numbers are in the tens of millions or merely a small handful.

On the timeline of human history, computer networks represent the greatest jump forward mankind has ever taken. Johannes Gutenberg's printing press(1400's) allowed us to mass produce books more efficiently than the slow traditional method of copying by hand. Guglielmo Marconi's work on the radio(late 19th century) made it possible for a person to be heard across the world as he is speaking. In high school Philo Farnsworth thought of the basic idea for television(1920's), adding the element of video to long distance broadcasting. During the middle of the 20th century, large computing machines owned by governments and universities were strictly used for scientific research such as weapons development and space exploration. By the end of that century, the computer became portable, affordable, and easily networkable. Any man, woman, or child in front of a computer can communicate with every other human being on the planet via this universal machine.

From the broad textbook view of the world, a limitless accessibility to the sheer information we hoard provides a catalyst for mankind's evolution. Any school on the globe can display a thousand books on math, science, and language from a basic level to a post graduate education. A borderless doctor can draw upon the medical knowledge of a few thousand years, the real time expertise of colleagues internationally, the pharmacy of an entire planet, and the technology of the wealthy. Modern warfare translates to a soldier on the battlefield knowing his enemy's exact weapons, GPS coordinates, detailed family tree, favorite foods, and complete ideology. Knowing how your neighbor in this global village lives, laughs, breathes, and cries may make it easier to choose civilization over extinction.

Darwin's theory amended to include computer use is prophecy, as technology becomes a favorable trait to possess. Besides the intrinsic advantages a computer provides as a workflow device, its greater asset is helping to share resources from one entity to another. Charities need generous donors, stores need frequent customers, companies need good workers, families need to bond, and lovers need each other. It is well known that countries, corporations, and clubs benefit from exchanging their raw materials, knowledge, and skills. Any group who chooses either isolation or ignorance deprive themselves of an entire planet's aid.

As the cost of technology drops over time it also helps to diminish the inequality of wealth. A laptop shared by a family living in an adobe hut may not be as fast as a corporate workstation at Adobe, but they are connected just the same. Encyclopedias on the internet are reachable whether you are at a wifi hotspot in a hotel outside Stuttgart, an investment banking office in Chicago, on a hardline connection at home in Lima, in any of a slew of internet cafes strewn throughout Seoul, or tapping away at 35,000 feet on a transatlantic flight. Weather reports showing average rainfall data, time lapse satellite footage, temperature ranges, and tidal forecasts are valuable to pretty much all the inhabitants of our blue heaven.

How is the universal "need for speed" satisfied across the frozen plains of the arctic wasteland? Asia's vast mountain ranges, Brazil's dense forests, and North Africa's hostile deserts present the same logistical challenge to getting decent streaming video. Daily life in the real world is a persistent battle for survival so therefore a trip into the virtual one is a fantastic luxury. Dealing with the world wide wait is another hardship to those users who have to share a computer and/or trek to where one is available.

Universal broadband is best served to the distant parts of the global village by skipping a generation of technology. Copper telephone lines crisscross the industrialized areas of the world like veins along the face of the earth's crust; it provided the internet's infant reach to individuals. Replacing them and expanding their current reach with a web of speedy fiber optic data lines to areas lacking phone service is already an ongoing project. Building slowly in parallel, cellular broadband is providing an alternate solution to those pesky geographic problems. Planting cell broadcast towers in remote corners of the Earth is akin to Neil and Buzz planting a flag on the moon.

"We come in peace and we bring you the internet," is the message from a number of companies landing here from the material world. Advertisements about enjoying a colorful iPad, a sleek new BMW, or a big screen TV are alien to the majority of the people currently on this planet. Despite the massive billboards and pervasive print ads across the skylines of Tokyo, Times Square, Bangalore, Dubai, Frankfurt, and formerly São Paulo, the consumerism lifestyle is still an elitist one. What they fail to tell you in the fine print is how much the cutting edge technology will cost from scratch.

A government or corporation has the collective means to plan, build, and maintain the infrastructure of society. The tribal resources of big brother allow us to have paved roads, clean running water, stable electricity, sanitation service, public schools, libraries, and internet access. In varying degrees your tax dollars support these utilities so you can benefit from their use for far cheaper than say digging your own well, collecting the water, and purifying it for your personal use. Depending on your latitude and longitude, you may be lucky enough to live where all those things are available because the people living there before you strove for it.

Suppose your situation is nowhere near utopia then it is imperative to get the engines of progress moving. Consider the Chinese proverb, "Give a man a fish and you feed him for a day. Teach him how to fish and you feed him for a lifetime." Let us modify the anecdote to incorporate feeding a person's mind with the resources of an internet enabled computer. Teaching a man, a woman, and all the children to use the computer is pivotal to improving their entire course of life. You can learn any real world skill, trade, or knowledge with the computer's infinite reach. The wisdom of village elders, teachers, and craftsmen across several cultures can be channeled to instill a sense of self sufficiency.

While the network of computers that comprises the internet has a tremendous capacity, the throughput is still finite. With a few billion people not yet online, the information superhighway has yet to hit a gridlock. Moving words and pictures across the surface of the world would probably be manageable even if every human being just suddenly had personal email access. Transforming the world's telephone array into a VOIP system would be an interesting half step into the future. Movies and television shows on demand along with video conferencing are enough to break the bank though. The amazing race boils down to engineers upgrading the structure of the internet faster than politicians can push their dramatic promises.

High speed routers and switches manage the data from node to node, computer to computer, like complex information pumps. Ideally data travels across the long distance as pulsed photons inside a fiber optic thread. These light pipes are bundled together in a stiff cord, carrying streams of data literally at the speed of light. Split from the major freeways of data traffic, countless capillaries of metal wiring shoot off from various Internet Service Providers to your home, school, or office. In the naked air, radio waves compete for recognition inside a specific frequency range in order to transmit precious web pages.

It is called "the last mile" by the planners who figure out how to get internet access into your anxious little hands. That final leg your downloaded song has to travel to get to your personal computer is the most troublesome one of all. Of course the net's main arteries and junctions are meticulously maintained by professionals whose lives revolve around preserving the flow of information. Your path is often beset by technical difficulties beyond your control, mounting upgrade costs every time technology has a slight change, and the occasional disruption by rogue pets and children.

As the number of computers in a network gets bigger, the overall system increases incrementally in complexity. Consider how raising a family is just a little trickier than simply being in a monogamous relationship. Managing a classroom well defines the skill of a teacher in encouraging thirty emerging personalities. If you have a hundred computers all linked together, potentially you could have the chaotic noise of everyone talking at the same time. On the internet, you have millions and millions.

Fortunately at this stage of the computer's evolution, most individual users are barely touching the surface of what their machines are capable of. The pipes of data transfer barely a trickle when you send an email or enter a line in your instant message window. Fetching a web page is a momentary data spike followed by a stagnant pause while we read the article, email, or snippets of feedback posted for your amusement. Generally speaking, our routine daily internet activities require a minimal amount of data traffic.

The old school paradigm of a server-client network model is going to be around to meet our needs for a very long time. It is a simple concept where you use your laptop or desktop to request files from a computer that has what you want. By pulling up a web page, it is like we are are stopping at the local MacDonald's to order the food we come to expect. A woman in front for you points to the menu and orders a Big Mac, you have a large order of french fries in mind, and the kid behind you is excitedly raving about a soft serve cone. As it is lunch time and a tour bus has just pulled up, a line of anxious customers is starting to form behind you. That single mother who casually stopped in to get a carton of milk for her five year old now has to wait in line a lot longer. So the next time you are stuck waiting for a web site to load, contemplate how many others are waiting in line.

Distributed computing, or uniting many computers to do a task, represents a more efficient way of spreading a chunk of information. From a user's point of view, peer to peer(discussed earlier in the mushroom chapter) is widely known for its key role in free file sharing. From a business perspective, it is an amazingly faster and more resource friendly way to broadcast things to your customers, clients, and fans. As a musician you can send promotional singles to everyone on your fan club mailing list without the cost of burning a disc, printing cover art, mailing materials, or postage. Government notices could be propagated to every internet enabled household via a regular desktop computer instead of an expensive series of servers. Network administrators in every sector can send software updates to thousands of client computers in a matter of hours rather than days or weeks. Anyone with a video camera and personal computer has the capability to put out his or her own channel for anyone in the world to watch.

In addition to being a mighty broadcasting tool, a slew of networked computers can invoke their collective power as a supercomputer. This colony of electric brains can take a massive problem and mathematically tackle it in pieces. Predicting global weather patterns, creating earthquake scenarios, mapping the human genome, and sifting through space noise for alien life are some of the projects on the table; you can install the BOINC software from http://setiathome.berkeley.edu/ to join SETI's joyful search for extraterrestrial intelligence . Donating your computer's idle time is a painless way to advance mankind's understanding of the universe without you personally having to wash beakers at a research laboratory.

On the blue sphere called Earth, it is easier to count the six billion inhabitants as already connected via the internet. The circumstances of dire poverty may delay their inevitable participation by decades but their digital voices should be noted. Take the current issues we are facing now in regard to the internet and multiply them by a factor of five to properly account for everyone. Consider the unrealized contribution of several billion anxious minds hungry for the promise of instant information and interaction. Also imagine about five times the online crime, censorship, information on you, energy usage, and electronic toxic waste.

An organization's web site is as prone to vandalism, burglary, and protests as its real life headquarters. A Denial of Service(DoS) attack which takes down a web server is akin to having a flood of homeless volunteers fill your restaurant and crowd around the entrance so no paying customers can even see the front door. A hack on your home page can result in online graffiti, much like someone walking right into your living room and tagging your walls. A quiet invasion into your server by computer ninjas may be just the thing to publicize your company's future marketing plans, hidden financial records, and those dirty little secrets regarding poor product safety.

Those crazy movies where the villain hacks into a computer network to set off a nuclear warhead, disrupt traffic, spy on you, erase your bank account, and take over the local television station have a disturbing grain of truth in them. From a technical point of view, all those things that happen to Will Smith's character in Enemy of the State are possible. The fourth Die Hard movie where the bad guy takes down the communications infrastructure of the United States of America by a series of calculated steps has a certain degree of plausibility. We may be a distant reality away from being enslaved by killer robots but George Orwell's theme of a omnipresent Big Brother is already here.

In the computer security community it is well known that no system is completely safe from someone trying to break in. As technology improves, it gets easier for people to get into places like the Pentagon, investment banks, universities, Microsoft, or whatever strikes their fancy. Although the average person might not ever consider doing such a thing, exploiting the weakness in some networks is as easy as following a recipe. Once a vulnerability is discovered by a clever hacker, he can easily write it down for someone else to try. In the vast maze of the internet, there are countless keys to your kingdom floating around for anyone to find.

Organized crime has to have its own IT division to protect themselves against the law, aggressive competitors, and mischievous explorers who wander into their servers. A street gang who crosses paths with a disgruntled hacker could have the DEA, FBI, and IRS breathing down their door all on the same day. Physical threats are intimidating to an individual and his family but retaliation on an organization just takes one computer. On the battlefield of cyberspace, it is a whole new set of rules, tools, weapons, and skills for every player to master.

Hate groups should beware of how loudly they proclaim their superiority over others as well. In real life there are many

167

places to meet in secret to reinforce your ideas with other like minded people. With an online presence, there is a truly unique form of democracy at play here. Whatever you have on your little web site is open to the entire universe of public opinion from left to right, rich to poor, black to white, gay to straight, and angry to loving. A few of the six billion others out there may not necessarily agree with what you have to say.

It is possible that the internet is the perfect place for each of the various factions of humanity to build their clone army. The religious will have their clusters of web sites proclaiming the righteousness of their particular God view. Science fiction groupies will have their various web rings celebrating Star Trek, Andromeda, Star Wars, Stargate, and Battlestar Gallactica. News junkies will cycle through multiple tabs on their browser, displaying the latest updates from both hemispheres as they are reported. Nationalists living at home or abroad can obsess about their holidays, language, foods, geography, and flag twenty four hours a day online. Sports fans can celebrate their favorite teams with collaborative videos, songs, conventions, and private parties.

The bombings, shootings, beatings, insults, and ignorance will continue in the virtual world for as long as people fight on earth. Cyberspace is not some utopia where people have a heightened sense of civility or kindness. This digital landscape is a pure reflection of the space that resides between our ears. Good and bad.

Of course certain knowledge is forbidden by the state. Depending on what country you happen to live in, the restrictions of censorship are heavier than others. Several of the governments who have known human rights violations also happen to be the ones who impose a ban on various servers outside their country. When you try to view any of a list of banned IP addresses, you are presented with a page not found error.

Although most users will be hampered by the digital blockade, anyone with a little bit of technical expertise can find a workaround. Alternate web sites abound with mirrors or similar applications to the ones which are officially banned. Seemingly innocuous forums will have links to the information you want using

168

revolving server addresses. Other sites will provide instructions on how to disguise your computer's IP address or how to tunnel to a private server. Much like the futile exercise of prohibition in the early part of the 20th Century, people will find ways to get what they they want.

In other parts of the world, access to the internet itself may be deemed sinful by the word of God. Religious leaders who have a freedom of data problem do not make it a priority to provide public places for free internet usage. Under the strict rule of ancient scripture, the proliferation of the internet is left to the interpretation of the controlling minority. To some fundamentalists, the internet exposes a faithful person to literally all the evils the devil can possibly promise.

The active condemnation of offensive materials stored, transmitted through, or even mentioned on a web site is often manifested in real life death threats. Instances of simple parody, reasoned critique, and juvenile humor might be too profane for members of a given belief. Scorn once reserved for obscure social activists is now open to the growing millions of the blogosphere.

By not citing specific countries, ideologies, or the names of some highly sensitive people hopefully that makes me less of a target. At the moment, the chief law of the land guarantees me the freedom of speech but only within our borders. Even so, the unwanted attention of a rabid stalker from a grassy knoll will ruin my day.

However, you may live in an area where these concerns are a daily aspect of your life. Your family or friends might have been threatened with physical harm as recently as today because of a point of view they voiced online. If you find yourself reading this with an eye over your shoulder, keep in mind we are living in very interesting times.

When you stop to ponder the logistics of control, the nature of the internet makes the proposition laughable at times. While the act of you expressing a particular opinion may be punishable by death, the idea itself cannot be destroyed. An incriminating video can be spread to every corner of the world in minutes, at will and in full color. A book or image can be saved like a dormant seed and passed from one person to another faster than a handshake or a sneeze. Suppressing the flow of information would be much easier without the dozens of types of cell phones, laptops, GPS devices, billboards, and other internet enabled electronics available.

To keep an individual trapped in a mental fishbowl may be impossible once she gets her hands on Pandora's personal computer. Assume every woman immersed in a patriarchal situation were to discover opportunities for merit based work, an avenue for unlimited education, and an open window to the world beyond the neighborhood. With the barriers of language fast becoming meaningless because of real time translation software, she would suddenly have the camaraderie of three billion sisters from every walk of life. My favorite curiosity would be to see how the ancient Greek play Lysistrata adapts itself to your local tribe courtesy of the internet.

In the shallow end of the pool of instantaneous information, you will also find personal trivia on the guy sitting next to you. His sexual preferences, annual income, alma mater, height, weight, date of birth, and primary vehicle are listed somewhere on the internet for you to find. A small biography can be compiled from public profiles on social networks, online resumes from job searches, personal web sites, comments from friends on photo galleries, and articles about a recent accomplishment. Anything from a getting mentioned in a brief newspaper blurb to accidentally walking into a stranger's photographic range will trigger an online web page of the event with your name and/or image. Bits of seemingly trivial information can be stitched together to compose a fairly thorough report of your life history.

For a few dollars more, you can graduate to obsessive celebrity stalker first class by accessing information databases aplenty. Formerly the private domain of aggressive marketing executives and underpaid county workers, these sources are now available to everyone via the internet. Things like property tax records, family trees, demographic shopping patterns, former addresses, criminal records, pending lawsuits, and Department of Motor Vehicle records can be obtained as easily as shopping on

Amazon. As long as you are a member of society on any continent, the paper trail now resides online for your convenience and misfortune.

People with paranoid delusions will also have a field day checking on the status of everyone they meet. Is the woman you are dating a compulsive liar when it comes to her inconsistent job history? Could the new coach/neighbor/teacher/pastor possibly be a registered sex offender? What does the guy you just met online mean when he says he has been married maybe a few times. Do you really know how many assets your future spouse may be hiding when he demands a prenuptial agreement? If the person is honest about how many partners he has had then looking up his medical history must fall into the acceptable snooping category, right?

This means your employer can cross examine your personal life with the same impunity. Those compromising pictures of you on vacation, at the office party, and in that magazine layout will be online for your boss to frown upon. The dozens of Dilbert clippings posted in everyone's cubicle are a subtle indicator that human resources is ferreting out your extracurricular escapades for management's permanent record on you. Anything from the past, your current after hours activities, and every little word on your fabled resume is up for scrutiny.

Privacy is becoming a precious commodity which neither the law, technology, or common decency can protect. Anyone in the public eye for more than their fifteen minutes may have permanently lost any claim to their anonymity. While the benefits of fame itself are fleeting, the nuisance of infamy will brand you forever.

An old timer in Scotland is in a bar talking to a young man.

The Old Man says, "Lad, look out there to the field. Do ya see that fence? Look how well it's built. I built that fence stone by stone with me own two hands. I piled it myself day after day for months."

"But do they call me McGreggor-the-Fence-Builder? Nooo..."

Then the old man gestured at the bar. "Look here at the

bar. Do ya see how smooth and just it is? I planed that surface down by me own achin' back. I carved that wood with me own hard labour, for eight days."

"But do they call me McGreggor-the-Bar-builder? Nooo..."

Then the old man points out the window. "Eh, Laddy, look out to sea...Do ya see that pier that stretches out as far as the eye can see? I built that pier with the sweat off me back. I nailed it board by board."

"But do they call me McGreggor-the-Pier-Builder? Nooo..."

Then the old man looks around nervously, trying to make sure no one is paying attention.

"But ya fuck one goat..."

Maybe it is the touch of schadenfreude, that delight in seeing someone suffer a dash of misery. My own theory rests on the sheer disbelief we cultivate when we witness any human accident. Not only are we shocked when we hear someone say the absurd, we are compelled to show someone else to make sure we have not imagined it. Thanks to the nature of the internet, we can pass that embarrassing video to our friends at the speed of light. To the chagrin of the poor schmuck who has been filmed, we can watch it over and over and *over* again.

After you have finished savoring the intimate accessibility of the global village, there are still the bodies to bury and the trash to dispose of. There is lots of waste to deal with in the brave new super fast internet enabled everything you want anytime world. The whole ignorance is bliss cliché is going to into epic fail mode because the earth is a small completely contained ecosystem. Shooting all that plastic packaging into the fiery heart of the sun is not an option.

Given all the lip service to recycling drink containers, newspapers, and motor oil, we predominantly utilize a primitive way of ending our electronics. We throw it away. There are a few cities around the globe that collect the dead plastic husks with the shiny glass faces in Green programs; we quietly assume they will not end in our groundwater. Unfortunately for the poor humans living in the

172

unluckier parts of the world, they get the seepage from both ends of a computer's life cycle. Tasty toxic metals will find their way into rivers, lakes, and streams as manufacturing byproducts and as they die off as junk. Nations lacking the economic forte to nurture bustling electronic gadget factories have the perverse luxury of sifting through the electronic remains of refuse pawned off by a richer country. It would truly be a landmark episode of Dirty Jobs to see impoverished scavengers wade through mountains of old computers to find random precious metal pieces for reclamation.

It may be a definitive complaint of my generation, but we never got the paperless office we were promised either. An arbitrary office picked from any old industrialized corner of the globe, probably uses ten times as many dead trees than a secretarial pool from yesteryear. We may manage more data, command more business, and actually be more efficient than any preceding decade. However we still cannot kick the paper habit no matter how devoted we are to almighty email. May the computer gods have mercy on your soul if you have a habit of printing out every message you receive.

Being a pulp fiber addict myself, it is most likely the subtle portability that seduces me. Lacking any electricity, my words still retain their form when they sit on the printed page. Until the other eighty percent of the world has a permanent connection to the world wide web, the book is still mightier than the you-know-what.

Getting back to the logistics of widespread computing for a moment, there is a big power problem which has yet to be addressed. Where do we get the electricity to fuel the next step in the international information revolution? Consider the adoption of additional personal computers coming online, work computers for new businesses being created, and more servers for the growing needs and numbers of web sites. Anyone familiar with a server farm already knows that when you have more than a handful of computers in a room, you also need to have the air conditioning constantly running to keep all of those sensitive CPUs happy. Right now the utility companies are slowly eating up the limited supply of fossil fuels on spaceship Earth thanks to us.

I sincerely think it is time for a unified push for solar

energy so us puny humans make it to the next few hundred years. While the solar panels on your laptop, clothes, or backpack or may be a clever option for the Eco-yuppie, I am referring to solar collectors on top of every building that faces the sun. Private industry needs to fund scientists to develop more efficient solar cells, governments should set an example on each of their buildings, and artists can work with the engineers to implement working designs with beauty. We are going to need a lot more juice than we have now to help several billion people, eagerly waiting in line, get themselves online.

Screwdrivers and Sunsets
12

In Ridley Scott's grim visions of the future, we are living on a heavily polluted planet where we use androids as slaves or we are being terrorized by frightening alien creatures with acidic blood. Thomas Malthus merely thought we would suffocate the Earth by sheer overpopulation. In Gene Roddenberry's positive view of the way things would work out, we would not only overcome racial differences, cure hunger, and befriend other worlds, we would also make it our mission to trek across the stars, freely exploring the universe. James Cameron's shiny metal assassins relentlessly return time and again on a unwavering mission to extinguish humanity. In a Transmetropolitan city expelled from the mind of Warren Ellis, we have a morally corrupt government, petty narcissism amplified to an art form, technology reeking of fantasy, the same inescapable age old human vices, and yet a feverish desire for truth. Whatever will be, will be but the computer's influence will undeniably color the world's fate.

What will the electric genie reveal about us?

Technology can be as mind consuming as alcohol, gambling, scrapbooking, drugs, Nascar, coffee, and chocolate; technically alcohol is a drug but the social rituals evoke an entirely different party. This modern addiction combines the shiny allure of designer jewelry with the human craving to fidget with buttons, levers, and switches. A touch of upgrade fever leaves your pocket book bare, clutters up your meager home, and often ruins your social skills by continuous neglect.

Deny it and you plunge yourself back into a shrinking primitive world. Ignore it and watch society around you grow smarter, faster, and more informed. Embrace it for the rush of cocaine consumerism, aka a lifestyle of shiny happy toys and a perpetual hunger for more. Loathe it if you will but you are still going to be surrounded by users, pushers, dealers, growers, and the government using it for their own means. Overdose on a surround sound home theatre parthenon with a classic Norad inspired multi-

175

monitor command center equipped with the latest gaming hardware for that deep dark antisocial trip into the abyss. As long as you take your internet in moderation, everything will be okay.

Unfortunately, your fondness for the latest gadget kind of opens you up to a whole new facet of crime. In the olden days, a petty thief would simply make off with your stuff and an off the shelf replacement would dry up those tears. When you lose an iPad, laptop, pda, cell phone, or personal media player though, your important data has disappeared too. Take a few minutes to ponder the loss of your personal information in this day and age. Those music files are probably backed up on your main computer as are the videos you pulled down from the internet. However, any laptop work files you poured your blood, sweat, and mouse tears into would now be in the hands of somebody else. In a worst case scenario, your address book and financial information would be totally exposed.

Muggers, car thieves, burglars, and hackers have a tendency to go after the easy targets because weak prey tastes the juiciest. The knowledge that everyone is vulnerable will either amplify your paranoia or lessen your fears about someone breaking into your computer. Consider for a moment that anyone who really wants to spy on you can do it more easily than ever. Access to the internet lets anyone have instructions on how to hack into a network, recipes for kitchen chemistry explosives, encryption algorithms for every security protocol you can think of, the basic principles to successful identity theft, and the psychological profile of a typical victim. Swimming in the same technological world as the so-called criminal element, levels the playing field by providing you with their same set of tools.

How do you bone up on every survival tactic, defense mechanism, counterstrike, online protection, and preventative strategy out there though? Any action you take uses up a certain amount of your personal resources to learn, maintain, and apply to your computer system. Given a certain snapshot in time, your solution may be the best one available to your knowledge. Threats to your computer are evolving in the wild like the mutating bacterias and viral strains of the real world; the antibiotics and drug therapies are becoming less effective in certain circumstances. Someone who has a grudge against you will devote the time to be more clever, persistent, and inevitably more dangerous than a graffiti minded kid.

Information itself is a fine art to master and a science to be reckoned with; this is where we interview our nearest marketing strategist, database programmer, translator, political lobbyist, or graphic designer for proof. It is an intangible necessity like the air we breathe but a strictly immeasurable abstract. As a piece of energy or mental food, it sustains our insatiable brains from cradle to grave. Consuming too little input makes our minds smaller and our perception of the world more narrow. A large variety of sensory data does help to fill in our framework of reality. The human mind is equipped to handle a tremendous amount of information, as long as we remember to pace ourselves.

Modern mammals consider information overload a natural measure of progress. From the hundreds of channels beaming down from a sky full of commercial satellites to this wild and crazy internet thing, our cups truly overfloweth. Even if we exclude the computer portal, we can easily spend a few years with the booty from a video store, public library, book store, newsstand, or video game aisle. Throw the old world wide wonder into the mix to see who can adapt to several programs layered on a glass screen for hours on end, every day for the rest of their lives. Each individual program will also have several pictures, paragraphs, and videos to select from plus a slew of unused icons, advertisements, and unknowns to reject. Every day you have to mentally sort everything by importance, filter for offensiveness, and flag for priority attention. When you hit an emotional trigger you have to stop the train, put out the fire, do some damage control, and get back on track.

When "too much" takes its toll, you will find yourself fighting for your life. The specter of creep is not nearly as immediate as a gun shot wound, stabbing, land mine, heart attack, car accident, stroke, or plane crash. When you are sucked into the computer world for long periods of time, it requires more and more of your attention to keep up with the gremlin inside. That invisible tether is so strong that it tends to make you skip meals, postpone people, stop exercising, trade sleep for coffee, rationalize the nutritional value of junk food, freeze you in place, and lock your gaze to the screen

177

(nothing beats a classic Nerf ball for an alertness check).

Suppose you are able to wrest yourself away from your computer long enough to attempt a balanced life. Your time crunch still needs you to have either a thirty hour day or an eight day week. After working overtime, burning daylight for a commute, figuring out what to eat, cleaning up after yourself, staying in fashion, dealing with the broken things, and nurturing your relationship, you have to squeeze in time to check your email. Everyone says computers improve your life by making things more efficient but they forget to mention the human ambition to do more once things get easier. We content ourselves with the low hanging fruits of life and as soon as we get a ladder, we strive to savor the best peaches on the tree.

There is never time enough because there is so much to read in this newfangled computer driven universe. Manuals, tutorials, updates, tips & tricks, and new product comparisons are a lot to keep up with just to use the damn machinery. This is the age of increasing download speeds, large monitors, and multiple displays. On screen, you must parse through several windows with additional unseen pages embedded underneath and scrolled over. Each time you jump to another web page or program interface, your eyes need to adjust to a different overall layout. Ignore those excuses because your personal reading speed is going to be the ultimate limiter on the information superhighway.

The pace of information will drench you like a summer monsoon, with a deluge of data raining down on your face. Unless you are already a twenty first century homo superior, you had better adapt yourself quickly. Anyone affluent and under thirty will have benefited from a childhood of the latest technologies; the test of wealth is not having expensive clothes but access to a home computer with broadband internet. It is neither the weak per Charles Darwin, nor the meek per the Bible's Psalms, but truly the geek who shall inherit the earth.

Anyone who lives as an Amish, Aborigine, indigenous, tribal, or likewise anachronistic soul will soon become museum characters. Picture a set of wax figures in native costumes complete with a typical meal, marriage ritual, religious ceremony, grooming habits, and birthing practices. On the summary description printed in large print black and white, you discover the enclaves where these rare peoples can still be found or where they once lived. They have remained off the grid and in the wild for as long as possible, in quiet little pockets of the world.

Finding a group of people untouched by a lone computer, cell phone, or any aspect of the internet itself is still quite easy. Any area where running water is a luxury will not be able to watch the idiocy, sadness, humor, hope, and wonder plastered on YouTube. A country pockmarked by land mines, mortars, rubble, mass graves, and militia does not debate the merits of Windows versus Mac versus Linux. In a village where women are sold as property, the shiny glass faced iPad stands as an artifact from outer space. No one at the RIAA or MPAA need worry about any bit of piracy nonsense from families too poor to feed themselves.

Whether by famine, fate, or fear, these tribes face a widening distance between them and the elite forces of the world; the minority of the wealthy one percent may well control global politics, financial markets, transportation, fuel, corporations, real estate, and possibly even the weather. It could take a span of years for an individual or a partial generation for a family, but the gap can be bridged by the wormhole of technology. This internet genie is powerful enough to bring the starving poor to the echelons of the upper class. By the same token, the doorway leads back to the fall of any empire who fails to utilize technology well. A single individual or small group can change the course of the planet with a laptop.

Good...bad....who is the one with the computer?

Besides the poor and uneducated, there are huge consequences in store for the rest of the non-technical inhabitants of Earth. America's middle class, Europe's working class, Africa's emerging class, and Asia's rural population are waking up to an entirely new day. Taxes are paid online, votes are being counted electronically, postage stamps are starting to lose their meaning, and airlines have stopped issuing tickets. Newspapers which once provided a universal means to inform the public, are slowly becoming an endangered species.

Lawyers, lawmakers, and judges are struggling to understand the implications of the rapidly changing paradigm, often without proper technical expertise; their positions afford them even less time in their schedule to fully comprehend Mr. Computer's effect on society. Add up the thousands of pages of legal code(state, federal, and international) for both the criminal and/or civil aspects of their specialty while keeping up with the ongoing cases shaping the living law. Everything from intellectual property and workers compensation to treason and homicide to marriage and divorce are all affected by the lens of technology. Evidence, testimony, warrants, subpoenas, and case files are each recorded, tracked, archived to some degree through a computer. When your attorney entrusts his personal notes, email correspondence, and financial records to a single on-site office server, the law firm should have a solid backup plan. Legislators who fail to maintain at least a conceptual understanding of the technology in their pocket, on their desk, and in the world are shortchanging their districts. Every judge from Small Claims Court up to the Supreme Court bears a greater societal responsibility to have a practical grasp of computers under their belt; it is their professional obligation to steer us toward a peaceful coexistence like the one embodied by the United Federation of Planets.

Doctors, nurses, and other intrepid medical practitioners who work a healing craft benefit from the productivity boost of a few cups of internet. In the eternal battle against the dark tower of bureaucratic paperwork, they need every advantage in the book. Any physician who does not have an active email account or follow online medical journals on a regular basis is one step away from laying on of hands and old fashioned snake oil. The success of human medicine depends on their free exchange of new discoveries and our understanding of their work. When the threads of electronic communication break down, the uninformed patient suffers from the treatable conditions of fear, ignorance, confusion, a sense of helplessness, and a lack of resources.

Teachers who do not keep up with technology are failing their students in a sadly ironic way; as facilitators of learning, they owe it to themselves to embrace computers as master tools of their craft. Ideally, technology can be integrated into every classroom and curriculum all through K-12 for each subject; they are the eventual addition which will supplant the traditional encyclopedias, chalkboard, overhead/film/slide projectors, wall world maps, and heavy textbooks in place now. Once society increases the funding for schools all across the planet and the district management becomes more efficient, the educators can teach with technology gusto; kids will have fingerprinting one day and digital image creation the next. When the convenience of calculators, laptops, search engines, internet tutors, and infinite libraries proliferate, the importance of a truly good education skyrockets. When the entire world is at a student's fingertips, our best teachers are (regardless of technology) the ones who have us think critically, develop a sense of character, examine many perspectives, regularly exercise our minds, and be personally open to new ideas.

For the average barista, sales clerk, truck/taxi/delivery driver, machinist, plumber, farmer, and unpaid housewife, technology is here to improve the way you do your job or get you the skills to find another one. The j-o-b is much less annoying when things are smoother, tasks are easier, equipment is safer, the process is better structured, and you are comforted by your favorite music. Everyone is empowered to find their own ways to improve their work by their employer during a downtime or by themselves at home. It is really for your own good to use your initiative to get what you need from your home internet machine. Investing in a computer can yield greater return than your pension, 401k, savings account, random bonus, or sucking up to your boss for a raise.

Regardless of your profession, managing your computer "stuff" is an enormous chore unto itself. Your bookmarks need to be sorted, your files kept orderly, your usernames/passwords kept safely, your data backed up regularly, and your software needs to be maintained; this does not even count the adventures with your computer hardware. How much time you devote to this aspect of life does not necessarily correspond to how proficient you are either. The unavoidable trio is really death, taxes, and managing your computer files.

Please enter a username, account ID, social security number, email address, or nickname. Click on the box to declare you are over eighteen years of age even though the only proof it needs is your word of honor. Type in one of your many super secret twenty character combination of mixed case letters and numbers passwords. Carefully retype it in on the next line to confirm it but erase them both to start over when you think you mistyped. Hit the Next, Submit, Go, Allow, right arrow, magnifying glass symbol, Search, Finish, Sign In, Sign Up, or Login button to proceed. In case you forget your password, you will be asked a very personal question which only you(and your spouse, best friend, mom, dad, sister, brother, or someone with access to your desk) know the answer. After you correctly answer the obscure trivia question about your life, your new temporary password will be sent to your backup email account in two shakes of a puppy's tail.

Single sign in solutions be damned, the typical person will have at least a dozen accounts to manage across several different types of online systems; my paranoia warns me about owning a universal electronic key with my tendency to misplace things. PIN numbers for your ATM, voice mail access, phone locks, apartment entry codes, security gates at work, network login codes, and retirement funds all need to be memorized for your protection. You probably have them written down on a note card, in a desk planner, on a worn out business card in your wallet, on that tattered desk blotter underneath your keyboard, or on the outdated wall calendar behind you. Of course the digital equivalent is keeping it on a purposely mislabeled text file on your hard drive, on a flash drive, as an attachment in an email to yourself, or on a web server. The deeply paranoid and/or clever can embed the information in an image, encode it to a secret file using open source military grade encryption, burn it to a compact disc to hide in a music collection, construct a fake crossword puzzle using it as clues, or mark up the corresponding letters in a copy of Lewis Carroll's *Through The Looking Glass*.

Getting in the front door is the easy part. Really. It is the file management which sets you in a tailspin, puts you in a tizzy, gets you all riled up, frustrates you to no end, and literally pumps up your blood pressure. After the post virginal exploration of where you find the basic programs on the computer, you will have to content with where the files go. Inevitably you will install new files, download them from the internet, copy them from another computer next to you, move them from one folder to another, save them to portable media, send them wirelessly, and then remember where you put them.

The science of search helps us to some degree but when we get a hundred or a few thousand results, it is not too useful when we want one specific thing. We can find wide swaths of information pretty easily but when it comes to the computer being a mind reader, it falls short. Expect a general answer when you ask a general question and very strict results when you ask for something specific; a generic search on a computer has as much potential comedy, frustration, and power as an episode with a certain fabled genie of the lamp. It is our responsibility to ask the right questions.

Whether you are searching for an email you received over four years ago, looking for a paper you worked on last night, scanning for a topic within a hideously long report, desperately trying to solve a problem using the default Help contents, or scouring the internet for a song you heard on the radio, your mental filters need to be top shape. The expression "trying to find a needle in a haystack" is like your search for missing file or needed answer. Normally you would be frustrated by the laborious digging through piles of straw or our situation with countless screens of information. A large magnet or fine metal detector would solve the literal problem whereas using a good search strategy helps us on the computer. Somewhere in your keyword choice being too general and too specific, you will get either the tidal wave or a fat ole goose egg. Sometimes if you type the exact question into the blank field, the all powerful computer gods will be able to respond with the answer you seek or offer suggestions on what it thinks you meant. If the information you are looking for is not currently available, you will have to have some patience and/or persistence.

Finding an answer through the computer, much like life *again*, is also a matter of determining if it is the right answer. Interpreting the truth is even more important in the intangible world of cyberspace because there are no boundaries of geography, schedule, taste, culture, or age appropriateness. Your life will be simpler when you start off with the cynic's creed to question everything you read, especially on the internet. Fortunately for all the propaganda, marketing, myths, falsehoods, hoaxes, half truths, and outright lies you come across, there is frequently a wonderful trail of crumbs to follow. Rabid comments by a ruthless online peanut gallery, neurotic bloggers with a chip on their shoulder, prominent debunking web sites, detailed money trail maps, and analysis from diligent experts are our saving grace.

Ultimately the real truth is up to us to find, accept, and understand. Reading an article online can easily be deceptive to the naive traveler; seeing it in a book or hearing it in person has no guarantees either. All the same basic rules of trust apply to the digital realm with one big caveat. Everything about the experience is cranked up a notch. There are more monsters and angels available to you than you would normally meet in your own little neighborhood. Your opportunities for temptation, salvation, mortification, education, elation, gestalt and perspective are magnified.

It is okay to feel a little overwhelmed.

Each of us would really just like the thing just to be simple. We want to have a computer which is easy to use. We would like it to last for a long time and not have to buy a new one every few years. It would be great if it did not to break down so much. A clear and easy to read troubleshooting manual would be a godsend too.

As inventions go, the personal computer has mainstreamed itself with an uncanny speed. Despite its complexity, the average person embraced it as readily as the bicycle, telephone, television, automobile, and community electricity. The motto, "Designed by engineers and scientists for engineers and scientists," does not inspire normal people to buy them in droves. Yet we do. Somehow these machines even motivate artisan painters to become animators, digital artists, graphic designers, user interface artists, video game architects, web designers, and logo specialists.

The everyman depends on the personal computer for communication(family, friends, and world), learning(school, news, and infinity), music, entertainment(games and videos), and managing money. He brushes his teeth in the morning, sits down to breakfast, and checks his email. She wakes up to the sound of internet radio, washes up, and reads the online newsfeed with her coffee. On the way to work, Jayson gets a two page briefing for his 9am meeting sent to him on his smartphone. In Miss Deborah's classroom, she teaches the geography of Europe using a holographic map projected in mid air. After dinner, your family gathers around a big screen to watch a rerun of a classic television show in a post nuclear internet age irony.

To maintain the ideal picture of your regular driving experience(commuter, race car, taxi, tractor, tank, plane, bicycle), there are certain things you do. The tires need to be inflated properly and be fresh enough to have a decent tread. Unless your vehicle is strictly people powered, your fuel should be clean and plentiful. Mechanically, there should be no loose, missing, mismatched, broken, or worn out parts. Keeping the interior vacuumed and the exterior washed improves our mood and hygiene, as well as prolong the life of the car. Tune ups are essential to keep everything running in tip top shape. Aftermarket upgrades also provide a higher level of customization for the functionality, look, and performance of your machine.

The same story goes for our personal computers.

We anthropomorphize them as trusted companions as much as we do with our beloved cars. To some extent, we learn to care for our trusty notebook so it faithfully gives us what we need. When we neglect it, we feel a wrath not unlike a petulant child or an abused animal. Coax and coddle it just right and it loves us back with the sights and sounds we want. Opening a new box of freshly packaged tech can be as thrilling as puppy day at the pet store. As time goes by, the electronic lifeform shows its age spots in the form of mental glitches and atrophying parts. Once it dies, the mourning period may be brief as long as you are able to preserve its data soul.

Even after all our fears, confusion, ignorance, and disdain for technology are dealt with in a reasonable way, we may still struggle with a lost innocence. Perfect health, long life, instant adaptation, endless amusement, and a fountain of endless knowledge are the gains in this trade off. At humanity's plateau we will have paradise woven into the fabric of existence by the seeds of today's computers. By then, we will know the answer to everything we care to question, see everything we dare to explore. One day however, the joy of discovery will disappear.

We have an unspoken millennium until that happens though. For you personally, it may already be true. It is not the part about Eden on earth but the lost sense of wonder. Whatever emotional scars, financial burdens, mental challenges, logistical hurdles, and physical obstacles you have to face to get online is worth it. Especially at this junction of history, we have the potential to shape the world with a small investment in ourselves. Lest you forget, these machines only have a supporting role to play.

Your personal computer or any piece of technology for that matter is just a tool. Like a basic screwdriver, it has a function. This is not to replace human interaction but to facilitate it. It can never take the place of the warmth of someone's touch or the simple unpredictable beauty of a sunset.

Glossary of Computer Metaphors

4chan... a place where chaos is your daddy
antivirus... your personal bodyguard
Alienware... a sports car brand of personal computer
Apple... a luxury brand(like BMW) of personal computer
augmented reality... seeing the world with a magic eye
avatar...an online version of you
Bing(.com)...to search for whatever you want, brought to you by Microsoft
BIOS... the waking up sequence of steps.
bit torrent... file sharing with your neighbors across the world
blog... an online journal
blue screen of death(BSOD)... your Windows(Tm) is having a stroke
bluetooth... the wireless signal for your earpiece
blu ray... movies that are too clear.
browser... the car you drive from web site to web site.
Cable modem... the internet box from your cable TV company
chat... real time online faceless chatter.
cloud... a bunch of computers working together in the sky
computer... a magic television typewriter tool
cpu... the brain.
Desktop... 1)a full sized computer , 2) the workspace on your screen cluttered with icons
download... bring it to you
drag and drop... move your stuff from here to there
driver...the instructions your computer need to understand what you just plugged into it
DSL.... the internet service from the phone company
ebook... a paperless book
Email... a digital letter
emoticon... shorthand for emotions
ethernet... the fat phone cable that connects your computer
facebook(.com)... the backyard bbq network
FAIL... an exclamation of utter incompetence or failure

file sharing... borrowing a movie, book, or music from your invisible neighbors

FIOS modem... the superfast internet box from the phone company

firewall... the bouncer at the club

firewire... a type of connector for video cameras

Flash drive... portable memory

flash memory card... portable memory for your camera

font... the way your letters and numbers look.

Forums... the online peanut galleries

Geek... someone obsessed with technology

gigabyte... almost a whole movie

google(.com)... to search for whatever you want

hacker... someone who breaks into your stuff

hard drive... the back pack for all your stuff.

Home page... the front door to your home on the internet

IM... an immediate digital shout

install... nail down the program to the computer

Internet... a library infinite with any book, movie, and song in any language at any time

IP address...the street address of a computer on the internet

keylogger... an invisible spy

kilobyte... a fraction of a paper

laptop... a computer you can fold and take with you.

linkedin(.com)... the office network

malware... an vandalism program which tends to be pervasive and annoying

megabyte... a minute of music or a few books

minimize... shrink the window

modem... the internet box(as in dsl modem, cable modem, or FIOS modem)

Motherboard... the nervous system.

Mouse... the puck which moves your pointer

Mp3... the common type of music file

MySpace(.com)... the local bar network

nerd... someone obsessed with information

nerdcore... what a nerd considers porn

netbook... a 9 or 10 inch laptop

newbie... a beginner

notebook.... another name for a laptop

office... a group of programs that obsoleted a typewriter, charts, and a slide projector

open source... any software that's FREE to download, use, and dissect

Operating system... the culture of a computer

phishing... fake emails or web pages designed to steal your personal information

photoshop... the program you use to doctor your pictures

pointer... your finger on the screen

portable apps... a compact version of your favorite program for your pocket flash drive.

Power strip... a row of electrical outlets on a stick

printer... the paper spewer

pr0n... naughty pictures

RAM...it represents the amount of things you can manage with your hands.

reddit(or)... a lover of narwhals and bacon

Router... the transmitter for your home internet signal

search engine... the "know it all"

server... the big computer[in the sky] feeding your computer.

social networking... the obsessive compulsive way to stay in touch with your friends.

Spam... junk email

spyware... snooping programs

streaming... a flow of music or video to your plate

surge protector... a power strip with a circuit breaker

tablet... a digital drawing pad

terabyte... hundreds of movies or thousands of songs

texting... real time online faceless chatter using your phone

touchpad... the finger pointing area

touchscreen... a TV that responds to your touch

Twitter... the online fortune cookie broadcaster

upload... give the file away

USB... the rectangular connector used for many of your computer cables

video card... a separate artsy brain which helps show videos
virtualware... a way to run another world
Virtual reality... a made up world
virus... a contagious blight that destroys data
VOIP... call your mom through your computer
webcam.... a computer camera
web page... a room in someone's house on the internet
web site... your home on the internet
white hat...a security superhero
wifi... a room's invisible internet signal.
WTF... an exclamation short form of "what the f*ck"
world wide web... another name for the internet

"The world isn't run by weapons anymore, or energy, or money. It's run by little ones and zeros, little bits of data. It's all just electrons."
-Cosmo from the movie Sneakers 1992

www.ingramcontent.com/pod-product-compliance
Lightning Source LLC
Chambersburg PA
CBHW051236050326
40689CB00007B/942